Nature-Nurture? *No*

Moving the Sciences of Variation and Heredity Beyond the Gaps

Peter J. Taylor

The Pumping Station
Arlington, MA

Published by The Pumping Station
61 Cleveland Street #2, Arlington, MA 02474-6935, USA
thepumpingstation.org

Library of Congress Cataloging-in-Publication Data

Nature-nurture? No : moving the sciences of variation and heredity beyond the gaps / Peter J. Taylor

- p. cm.
- Includes bibliographical references.

Library of Congress Control Number: 2014910853

ISBN 978-0-9849216-4-5 (pbk)   978-0-9849216-5-2 (cloth)

Revised October 2014
Printed by Lightning Source in **Gill Sans** and Adobe Garamond Pro
Available in digital form as a pdf from the Publisher, http://bit.ly/NNN2014

Table of Contents

CODA

Dedication

To the late Don E. Byth, who helped show me the possibilities and limits of plant breeding—I wish we had had time together to talk over the puzzles and gaps in this book—and to Richard Lewontin, a model for connecting genetics, agriculture, and politics.

Acknowledgements

This book arose from one strand of research supported by the National Science Foundation under grants SES-0327696 and 0634744. In this work I have benefited, either while a guest at their institution or in joint sessions at conferences, from conversations with David Barker, Yoav Ben-Shlomo, Werner Callebaut, George Davey Smith, Ian Delacy, Bill Dickens, Steve Downes, Jim Flynn, Jonathan Kaplan, Ken Kendler, Barbara Kimmelman, Diana Kuh, John Lynch, Aaron Panofsky, Diane Paul, Ken Schaffner, and Eric Turkheimer. A remark of Ann Blum led me to see that turning my work on nature-nurture science into a book could clear space for the other strands of the original research, which are touched on in Items I and J. The comments of Anne Fausto-Sterling, Richard Lewontin, Steve Orzack, Susan Oyama, Diane Paul, James Simpson, Hamish Spencer, members of the Genetics and Society Working Group, and anonymous reviewers of previous articles helped in the revision process, as did editorial assistance from Rebecca May and Maria Nardi. Christine Luiggi designed the cover based on artwork from James Steinberg (www.james-steinberg.com). Sections of the text are adapted from previously published articles: Taylor (2001, 2006a, b, 2007, 2008a, b, 2009a, 2010, 2012).

Introduction

Almost every day we hear that some trait "has a strong genetic basis" or "of course it is a combination of genes and environment, but the hereditary component is sizeable." To say *No* to Nature-Nurture is to reject this relative weighting of heredity and environment. Such weighting derives from researchers in the relevant scientific fields partitioning the variation in the trait into a heritability fraction and other components. This book shows that this partitioning provides little clear or useful information about the genetic and environmental influences.

Saying *No* is a positive move. When we put the relative weighting of nature and nurture behind us, interesting scientific and policy questions about heredity and variation can be brought into focus, reframed, and taken in new directions. Or, to be more precise, interesting questions follow from assembling a toolbox of conceptual themes that allow us to move away from the persistent formulation *How much is nature, how much is nurture?*

The conceptual themes advanced in this book emerged from puzzling over the positions and propositions of others that did not, for me, fit together. I hope readers appreciate the coherence of the picture I paint, but, even more, that they become engaged in fresh directions of puzzle posing and probing. After all, to move beyond the *gaps* I identify in the study of variation and heredity requires a wide range of inquiries from people in many different areas: in various sciences, from quantitative, behavioral, and molecular genetics to epidemiology and agricultural breeding; in history, philosophy, sociology, and politics of the life and social sciences; and in engagement of the public in discussion of developments in science.

*   *   *   *   *

Readers may already be puzzled: Why nature *versus* nurture? Surely it must be nature *and* nurture—we all know that traits develop over time through the interaction of the organism with genetic or hereditary and environmental or social influences. Indeed, the modern science of epigenetics, building on ever-increasing information about DNA sequences and how genes function, now shows us how chemicals from outside the cell can modify the activity of genes for the rest of an organism's life and sometimes even into subsequent generations. Moreover, since the late 1800s—well before advances in molecular biology—developmental biologists have been studying the mechanisms through which a single cell divides into multiple cell types and gets arranged into tissues, organs, and the organism's overall form (Gilbert 2013). The nature versus nurture in this book is not, however, a matter of development of traits over time. A sketch of the history and current state of the study of heredity, development, and variation is needed to set the scene.

When it is said that a person resembles their parent, two aspects of heredity—the transmission of traits to offspring—are being raised: how does an offspring *develop* to have the trait in question at all, e.g., its eye color, and how does the outcome of the development at some point in the lifespan *differ* from that person to the rest of the family or population. Development and difference do not necessarily require separate lines of research. Biologists often study abnormalities in order to gain insight about typical processes of development. In Swyer syndrome, for example, a child with XY chromosomes has female, not male, external genitalia. The influence of estrogen and progesterone on development is illuminated by the finding that, if these hormones are administered at puberty, breasts can develop and regular menstruation can occur (Michala et al. 2008). Variation under a more or less normal range of conditions can also help biologists understand typical development. The age at which a baby comes to walk, for example, varies according to whether it sleeps on its back or stomach, has been swaddled or carried around in a sling, and so on, and this variation has been related to the timing and degree to which the baby uses its upper body muscles (Fausto-Sterling 2014).

Difference can, however, be studied without providing much insight about development. For example, the eyes of fruit flies, normally red, are sometimes white. Biologists identified the location on the chromosomes that corresponds to the white-eye mutation early in the 1900s and later investigated the pigment-formation metabolic pathway and the enzymes involved as fruit fly eyes develop the normal or mutant color. Such knowledge says little about how eyes develop. Even on the narrower question of how eyes get to have color, a lot has to be known already about the development of the eye as a whole to make sense of how and when during the fly's development the enzyme produces color in the eye.

The study of difference separate from development—albeit with a promise of later returning to issues of development—characterized the emergence of the field of genetics (Sapp 1983). This approach has been very productive.

Advances during the twentieth century include the cataloging of mutant genes and their location on chromosomes; associating different genes with different enzymes (molecules that modulate biochemical interactions); identifying the chemical basis of genes—DNA—and the mechanisms of replication and mutation; revealing, then manipulating, the processes by which genes influence traits first in viruses and single-celled bacteria, then in complex, multicellular organisms; mapping the specific DNA sequence of entire organisms; comparing sequences among taxonomic groups; and tracing where and when in development specific genes are active.

No catalog or database of genes and DNA variants for any organism remotely resembles a *literal* "blueprint" or "program" for its development. Nevertheless, with ever-improving knowledge about genes and technologies to manipulate them, the field of genetics is now involved, not only in explaining how one organism differs in a trait from another, but also in illuminating various steps in how an offspring develops to have the trait at all. When the use of genetic knowledge and technologies is seen as a productive strategy to work towards teasing open the complexities of development, the blueprint and program metaphors can be put aside; there is less need to depict DNA as the master molecule determining the traits an organism has and to assert the dominance of nature over nurture. The study of genes in development can then take researchers in many directions (Gilbert 2013). Some might continue to focus on the location of genes on the chromosomes, the mechanisms of their coding for enzymes, and how to engineer their action in new ways. Others might want to use knowledge of which genes are functioning and under which cues—from inside the organism or outside—to gain insight about mechanisms by which traits develop. Yet others might want to continue, in the long tradition of developmental biology, investigating how cells become arranged into tissues, organs, and the organism's overall form, studying biophysical processes as well as the biochemical pathways linked to genes.

The study of difference separate from development also underlies a quite different branch of research on heredity. Very early on in the history of genetics, researchers such as Fisher and Wright saw the need to reconcile the *discreteness* of traits related to mutant genes with *continuous* variation in many observable traits, especially traits in agriculture of economic interest such as yield of plant and animal varieties or breeds. Discrete variants not clearly related to pedigrees of inheritance of a mutant gene also needed to be addressed.

Using models of theoretical genes—idealized "genes," not those later mapped and probed—and assuming that more of these genes are shared among relatives than in the population as a whole, data on the variation for a trait in a specific group or population could be analyzed so as to provide predictions of changes in the average value of the trait under selective breeding. The same methods of data analysis also partitioned the variation into components related to differences between the agricultural varieties, between the locations in which they were grown or raised, between specific variety-location combinations, and unsystematic variation. To choose a simple example, if the between-variety fraction of the variation—or *heritability* as this component came to be called—were zero, there would be no change expected under selective breeding.

Analysis of variation in ways that take genealogical relatedness into account without identifying inheritance of specific genes—the field of quantitative genetics—was soon extended from agricultural and laboratory breeding to analysis of variation in *human* traits, especially behavioral or psychological measures such as IQ test scores—the field of behavioral genetics. The emphasis here was not on selective breeding yet variation was still partitioned into heritability and other components. (This book often uses "heritability studies" as a non-technical synonym for the variation-partitioning aspect of quantitative and behavioral genetics.) The usual shorthand for the between-variety and between-locations components of variation is "genetic" and "environmental." Although these adjectives invite confusion (as discussed in Parts I and II), their acceptance is not surprising given that partitioning of trait variation has drawn from and fed into debates that pre-date heritability studies—indeed pre-date the field of genetics—about just how much a trait is nature *versus* how much is nurture. We all know that nutrition, care, and social interaction are needed for offspring to grow and thrive, but how much influence do we expect the environment or society to have on the eventual outcome—be it adult height or IQ test score—that results from that development?

*   *   *   *   *

The history of scientific and policy debates around this last question is long and *continuing*. How can the persistence of these debates be squared with the view that everyone accepts that any trait results from the interaction of nature *and* nurture? Delving into that puzzle led to others. The need to reject the partitioning of variation into heritability and other components—to say no to nature versus nurture—was not a position *directing* my inquiries, but a *conclusion* that emerged from the process of shaping pieces that do fit together.

The hard-to-reconcile pieces include not only the commonsense proposition that both genetic and environmental influences are always involved and the persistence of popular and scientific claims about nature versus nurture. What, in addition, is to be made of the divergent angles of critical commentary on nature-nurture debates? Commentators sometimes express skepticism about heritability estimates from studies of human twins, pointing to questionable assumptions about the upbringing of twins or once-important data sets being discredited. Sometimes they accept the idea of heritability, but remind us that "genetic" does not mean unchangeable—especially in relation to differences across groups in the average value of any given trait. Sometimes they worry that heritability values might be derived which are high and sound. (Parens and Chapman 2006 provide an overview of these positions. Stewart 1979 is an example of the worrying; see also Dickens and Flynn 2001.)

Moreover, as a counterpoint to critical commentary, researchers in human behavioral genetics consider that their field has responded to past criticisms, has moved on and assembled new sets of data, and is now employing powerful techniques to analyze those data. (Panofsky 2014 examines the history of these developments.) Plomin and Asbury (2006, 87), for example, present findings they consider well established on the heritability of traits and the size of the influences of *shared versus nonshared environment.* They point to the "flood of molecular genetic research whose goal is to identify the specific DNA sequences responsible for genetic influence on common behavioral disorders such as mental illness and on complex behavioral dimensions such as personality." Their conclusion is that those who resist genetic explanations of behavior are "sticking [their] heads in the sand"—"nothing [is] to be gained…by pretending that genetic differences do not exist" (2006, 96).

To make sense of these various pieces—to create a coherent picture—this book's key move is to distill issues into eight conceptual and methodological *gaps* that need attention (summarized in Table Intro.1 and presented in Item B). Some gaps should be kept open; others should be bridged—or the difficulty of doing so should be conceded. *Previous researchers and commentators have either not acknowledged all the gaps, not developed the appropriate responses, or not consistently sustained those responses.* When the gaps are considered as a whole, the limitations of the relevant sciences become clear. Quantitative and behavioral genetics do not provide a reliable basis for genetic research that seeks to identify the molecular variants associated with trait variation, for asserting that genetic differences in many traits come, over people's lifetimes, to eclipse environmental differences, for believing that the search for environmental influences and corresponding social policies is unwarranted, or for sociological research that focuses on differences in the experiences of members of the same family. Almost no grounds are left for remaining interested in the relative weighting of heredity and environment.

In summary, genes *and* environmental factors come into play along the pathways of an organism's development over time, but the mechanisms and details of development are not examined here. The nature versus nurture focused on in this book derives from the analysis of variation in traits. Analysis of variation, even when studied separately from development: Yes. Assigning a relative size to genetic and environmental influences on that variation: No.

*   *   *   *   *

A terminological note: "Factor" is used throughout this book in a non-technical sense, referring simply to something whose presence or absence can be observed or whose level can be measured—a quality that is emphasized in some places in the text by adding the adjective *measurable*. For any given trait, the factors of interest are those associated with variation in the trait between groups of individuals, but the *causal* quality of a factor is a secondary matter. Measurable genetic factors include the presence or absence of variants (alleles) at a specific place (locus) on a chromosome, repeated DNA sequences, reversed sections of chromosomes, and so on. Measurable environmental factors can range widely, say, from fertilizer application per hectare of crop, to average daily intake of calories, to degree of maltreatment that people experienced as children.

*   *   *   *   *

Readers ready for a step-by-step presentation of the gaps and their implications are welcome to fast-forward to Item B. From experience, however, I need to gain the attention of potential readers with various preconceptions—people who think that everything significant has already been said, or feel that technical expertise must be required to appreciate and resolve any further controversies; people who are not troubled by the different angles of critical commentary or the confident pronouncements from human behavioral geneticists. Item A is inserted, therefore, to introduce three significant puzzles *that have not been resolved by past debates.* Even then, if experience again is any guide, people intrigued by the three puzzles may be disinclined to read further given what the italicized assertion about gaps (at the bottom of page 5) implies. That is, despite decades of debate among methodologically sophisticated scholars, some fundamental problems in the relevant sciences have been overlooked. This claim may, understandably, seem implausible. Let me, therefore, sketch the background that allows me to see the study of heredity and variation differently from the many researchers, philosophers of science, and other commentators who have contributed to heritability studies and nature-nurture debates.

My initial research position in the mid-1970s was in plant breeding on a project to make sense of the ways different varieties of a crop plant varied in their responses across a range of locations. I had to analyze data from large trials, such as one in which 49 wheat varieties were grown in each of 63 locations around the world. A first step in the data analysis was to partition the variation observed in a given trait, say, yield, into components related to the averages or means of the varieties (across all 63 locations), the means of the locations (across all 49 varieties), and so on. (Indeed, agricultural breeding was where partitioning of variation and measuring heritability originated.)

The challenge was then to use the partitioning of variation to hypothesize what it was about any variety that led to its pattern of response across locations and what it was about any location that led to the responses of any variety in that location compared to others. Knowing what aspects of, say, the variety's pedigree or the environmental conditions contributed to the yield of that variety in that location could then inform subsequent breeding or cultivation decisions.

Such hypothesis generation was never easy, even though we had large and complete data sets to work from as well as a lot of background information about the varieties and locations. By implication, the limits to hypothesizing about genetic and environmental factors must be even tougher when researchers partition variation for *human* traits. In human studies any genetically-defined type is, at best, replicated twice—as identical twins—and different genetic types cannot be systematically raised across the same range of "locations"—families, socio-economic status, and so on. Appreciating this difference opens up one of the three puzzles of Item A, which concerns the potential for confusion that arises when methods and terms are carried over from controlled situations of agricultural and laboratory breeding to analysis of human variation, where control and replication are difficult.

Fast forward to a decade ago: The center of my research had moved to the social studies of science, where the topics I have chosen make use of my background in data analysis and modeling. I had a research grant to study researchers who use *observational* data (as against data derived from *experiments*) to investigate the complexity of biological and social factors that influence the development of any given trait over the human life course (Taylor 2004). The attempt of Dickens and Flynn (2001) to resolve what they called the IQ paradox especially intrigued me. Flynn had noticed large generation-to-generation advances in IQ test scores and found this result hard to reconcile with the trait having high heritability—another of the puzzles in Item A. Dickens helped him develop models consistent with both facts. I had some reservations about the models, which I shared with Dickens; he generously spelled out his thinking in response. This led me to convey further reservations. In the course of the exchange that ensued I found myself digging deeper into the conceptual foundations of heritability estimation and partitioning of variation. In order to present a picture that differed from what Dickens, Flynn, and others treated as unobjectionable, I was explicating first principles, not disputing details of specialized models, mathematics, or data sets. In doing so, I was making extensive use of perspectives and examples from the earlier plant breeding project. A particular aspect of that research led me to raise what is now the remaining puzzle in Item A, namely, the possibility that the genetic and environmental factors underlying a trait are heterogeneous, differing from one set of relatives to the next or from one environment to the next (Taylor 2006a,b, 2007).

This possibility of *underlying heterogeneity* had not, I found, been identified as a significant methodological concern by quantitative geneticists or by critical commentators on heritability research (e.g., Downes 2004 and references therein). Exploring the implications of heterogeneity more generally became central to my puzzle posing and probing to make sense of problems and possibilities in the sciences of heredity and variation. Moreover, as will be evident, my current efforts to make sense of research on the complexity of biological and social factors over the human life course are characterized by attention to the *concepts* more than *social context* of the research.

*   *   *   *   *

Roadmaps: Part I begins with the three puzzles (Item A), followed by the presentation of the eight gaps (Item B), which are then used to revisit the puzzles, each appearing in a new light with several new puzzles raised (Item C). The "Short Critique" made up by these three items leaves, as noted earlier, almost no grounds for remaining interested in the relative weighting of heredity and environment for human traits. Table Intro.1 (on pages 10 and 11) summarizes the gaps, as well as their connections to the initial puzzles and to background and implications developed in subsequent items.

The Introductions to Parts II and III provide more detailed roadmaps of what follows, as do the italicized summaries at the start of each Item. At this point, a brief overview: Part II amplifies the conclusion of Part I by exploring, on a more technical level, what heritability studies would look like if the second gap—the distinction between fractions of variation in a trait and measurable genetic factors underlying the trait—were sustained consistently. The exploration begins by introducing models that, unlike those of classical quantitative genetics, do not refer to theoretical genes (Item D). Using these *gene-free formulations,* the implications of alternatives to three standard assumptions of quantitative genetics are teased out (Item E). The highly circumscribed sense in which partitioning of variation sheds light on causal factors is identified as *rerun predictability* (Item F). Whether or not it is necessary to address our limited knowledge about the causal dynamics of a certain phenomenon is connected to the actions possible or proposed on the basis of what is known (Item G). Finally, some well-known accounts are reworked to show that, although it is by no means a new point to draw attention to the gap between analysis of traits and identifying relevant underlying factors, the distinction has not always been upheld, even by critical commentators (Item H).

Part III turns to research on the specific and measurable genetic and environmental factors underlying variation in human traits—research that makes no reference to a trait's heritability or the other fractions of the total variation. Two particular themes—the possibility of underlying heterogeneity and causal claims being grounded in practice—carry over from Parts I and II, helping to illuminate selected research programs in behavior, health, and epidemiology that leave heritability studies behind. Some of these programs still address genes and environment (Item I); others are discussed for their potential to address, not suppress, heterogeneity (Item J). Difficulties in

Table Intro.1 Eight gaps in analyzing and interpreting heritability

Puzzles (Items A, C)	Gaps (Item B)	
Puzzle #3— Transfer from agricultural to human studies —motivates the examination of all the gaps.	1. *Terminology*	Key terms have multiple meanings that are distinct.
	2. *Patterns versus factors*	Statistical patterns (e.g., variation for a trait partitioned into components) are distinct from measurable underlying factors.
Puzzle #1—IQ paradox—gets partly resolved by attention to gaps 2 and 5	3. *Translation to hypotheses*	Translation from statistical analyses to hypotheses about measurable factors is difficult.
	4. *Predictions*	Predictions based on extrapolations from existing patterns of variation may not match observed outcomes.
	5. *Unreliable estimates*	Partitioning of variation, especially in human studies, does not reliably estimate the intended quantities.
Puzzle #2— Underlying heterogeneity— feeds into gap 6 More puzzles follow, especially from gaps 6 & 8	6. *Translation to hypotheses (in light of heterogeneity)*	Translation from statistical analyses to hypotheses about the measurable factors is even more difficult in light of the possible heterogeneity of underlying genetic or environmental factors.
	7. *Causal influences*	Many steps lie between the analysis of observed traits and interventions based on well-founded claims about the causal influence of genetic and environmental factors.
	8. *Within to between groups*	Explanation of variation within groups does not translate to explanation of differences among groups.

Appropriate Responses (Item B)	Implications (and Background) (Items D-K & book as a whole)
Meanings need to be kept distinct, for which terminological changes can help.	
Needs to be highlighted and kept open.	Blurring this distinction leads to interpretations of components of variation that are unjustified -> Say *no* to nature-nurture (NNN). Critiques of heritability studies often abet this blurring (Item H).
The steps and conditions to bridge (or circumvent) this gap warrant attention.	Heritability studies are unreliable basis for molecular or sociological investigations of traits (NNN). Restrict attention to relatives (-> Item K).
Compensate for the discrepancies (especially if any actions depend on the predictions).	Causal claims circumscribed as rerun predictability (Item F), which make sense in relation to the actions based on the basis of what is known (Item G), a theme informing Item I.
Collect data sets needed to remedy the gap. Acknowledge when this is difficult.	Further unjustified interpretations of components of variation follow (NNN). (Demonstrated by examining alternatives to three standard assumptions, Item E, aided by gene-free quantitative genetics, Item D.)
The steps and conditions to bridge this gap—or to circumvent it—warrant attention.	Even less reliable basis for molecular or sociological investigations of traits (NNN). Pay attention to possible heterogeneity of underlying factors in fields other than heritability studies, e.g., epidemiology and social sciences (Items I & J).
Recognize that estimates of heritability and other fractions of the variation are of even more limited utility than gaps 3–6 indicate.	
Recognize that this gap is firm and its implications are deep.	

identifying causally relevant genetic variants underlying human variation are placed in a new light; further puzzles emerge—Moving beyond *no* on nature-nurture leads us to *maybe* for genetic and environmental factors. Indeed, Part III presents the possibilities and problems of the various programs of research in the spirit of unfinished and open inquiry, aiming to encourage readers to explore the book's conceptual themes in their investigations and discussions of heredity and variation.

A Coda (Item K) develops a suggestion—again offered as an issue to be further expanded on by others—that the conflation of family and population contributes to the difficulty of leaving nature-nurture behind.

*   *   *   *   *

Overlays to the roadmaps: In order to orient readers and distinguish my contribution from others, let me preview several features of the book's expository approach:

a. An *intentional and forward-looking focus on "classical" methods of heritability studies and the partitioning of variation*. Despite the conventional wisdom that "[r]esearch into the genetics of complex traits has moved from the estimation of genetic variance in populations"—what we might call *classical* heritability studies—"to the detection and identification of variants that are associated with or directly cause variation" (Visscher et al. 2007), it is not until Part III that I discuss research on the molecular variation associated with trait variation. I want first to draw attention to significant problems in classical heritability studies that have not been widely recognized or resolved. Doing so allows researchers to then assess whether similar or analogous oversights limit the level of progress that is possible with the newer methods.

b. The picture that results once the gaps are recognized *incorporates, but is not limited to, many critical points that have been made before*. Such points include not only the critical commentaries already mentioned, such as "genetic" does not mean unchangeable, but also the following: estimates in heritability studies depend on the specific population and situations for which the data were collected; estimates of the degree of genetic and environmental influence are not helpful in identifying the specific factors that make up that influence; and explanations that apply within groups cannot be extrapolated to apply to differences between groups. Past discussions of these points, whether by critics of heritability studies or by supporters, do not, however, address the full set of eight gaps adequately. Notably, I am dissatisfied with the account in Lewontin (1974a), "The Analysis of Variance and Analysis of Causes." Issues, such as underlying heterogeneity, are examined that have not been raised in the many critiques that have followed Lewontin's lead. (As an entry point to Lewontinian critiques, see the sample listed in Tabery 2009, which is a review of Sesardic's 2005 counter-critique.)

c. Although the book is radical—it calls into question what years of peer-reviewed research and associated commentary have put into place or taken as given—the *dominant style is pedagogical or didactic, not disputatious*. In places I identify shortcomings in the work of other authors, but my main impetus is to lay out "foundations" that, if built on, avoid the problems and confusions that beset heritability studies and their interpretation. This expository choice follows, in part, from my experience that dissection of the work of others is not rewarding when, as is often the case, understandings about the gaps are not shared. In this light, readers who might expect me to discuss researcher X or commentator Y should ask, for example, whether X and Y accept the interpretation of heritability as the "contribution of genetic differences to observed differences among individuals." (The quote is from Plomin et al. [1997, 83], but the interpretation is widespread.) To think that components of variation in individuals' traits provide insight about the genetic and environmental factors underlying variation in those traits is to discount the second gap. Meaningful discussion of the rest of the gaps is hard to have with researchers and commentators whose work relies on that idea.

d. *Puzzles* are used to draw readers into the process of identifying the eight gaps and formulating conceptual themes that allow us to move beyond those gaps. The three puzzles already mentioned lead off in Item A; further puzzles emerge—but are by no means all resolved—as the book unfolds.

e. *Data analysis* is the point of entry and central concern. This emphasis does not require readers to be well versed in statistical methods. However, readers need to put aside ideas and speculation about what genes can do or what genes can make organisms (humans included) do *unless (or until) those ideas can be assessed by some reliable method of data analysis*. The idea, for example, that genes elicit matching environments—nature *shapes* nurture—might sound plausible at first, but one has to ask how an association with *genes* would be shown through analysis of variation in observed *traits* in heritability studies.

f. The eight gaps apply to heritability studies for *traits in all species*, not only in humans. It is around human heritability studies that debate has been most active, but it is often helpful to consider methods of studying variation in animals or plants—the original domain of heritability studies. What agricultural studies can and cannot establish sheds light on what is involved or implied in analyses of data from humans. At the same time, questions also get raised about the reliability of heritability studies in selective breeding of agricultural and laboratory species even when some of the gaps can be bridged.

g. The *issues are conceptual, not technical or empirical.* Data analysis is the central concern, but no data, equations, or mathematical symbols are needed to present the Short Critique of Part I. The issues are expressed in terms accessible to non-specialists, aided when needed by entries in the glossary. At the same time, the sequence of gaps is intended to challenge researchers in the social and life sciences as well as other specialists and commentators. Of course, some of these commentators have made explicit and implicit responses to the gaps. However, when I acknowledge these responses, I minimize or separate out the technical details and refer readers to relevant items later in Parts II and III and in four Appendices. Moreover, although Part II involves equations and technical detail, conceptual themes are developed (as mentioned earlier: heterogeneity and causal claims being grounded in practice). It is these themes, more than the technical detail, that carry over to Part III.

h. Inquiry into *the social context of research is not the focus.* My contribution supplements historical and sociological work that makes sense of the persistence of popular and scientific claims about nature versus nurture, that examines the social dynamics of the scientific fields and the wider arenas in which their results are given meaning (Paul 1998, Panofsky 2014). As evident in Items C and K, close examination of concepts and methods within any given field of natural or social science—which is the focus of this book—can raise questions about the diverse social influences shaping the field (Taylor 2009b). I do understand that attention to social dynamics is important for the reception of a non-standard account such as mine (Taylor 2011), so the intentionally delimited scope of this book may well get rethought as responses of audiences or the issues raised by two books appearing at the same time as this one (Panofsky 2014, Tabery 2014) move me to refine my arguments and to experiment with different ways of reaching readers. To this end, *The Pumping Station*, with its flexible publishing mode, was chosen to expedite the initial publication of the book and make it straightforward to release revised editions.

i. The last feature of the exposition to note is the *cultivation of further inquiry*. My hope, as indicated at the start, is that others take up the conceptual themes and explore the puzzles or questions raised, leading to new contributions to discussions of heredity, variation, heterogeneity, and the social influences shaping research on those topics. For meaningful discussion and progress, appropriate responses to each of the eight gaps are a precondition. In that vein, it is time to move to Part I, "Puzzles, Gaps: A Short Critique."

PART I

PUZZLES, GAPS:
A SHORT CRITIQUE

Item A. Three puzzles not resolved by past debates about heritability

In which three puzzles, which have not been resolved by past debates, are introduced. The puzzles challenge the assumption that everything of significance related to heritability has already been addressed, or that technical expertise is required to understand any remaining issues.

Puzzle 1. The two-part argument and the IQ paradox

Flynn (1994) has pointed to large gains in average IQ test score between generations (now called the "Flynn effect"). No environmental factor, or composite of factors, such as diet or education level, has been shown to be strongly associated with these generational differences. At the same time, according to the current consensus, heritability of IQ test scores is high (Neisser et al. 1996, but see Turkheimer et al. 2003, Nisbett et al. 2012). Persistent large differences in average IQ test score also exist between racial groups (but with recent decreases, Nisbett et al. 2012). As with the generational differences, no environmental factor, or composite of factors, has been strongly associated with racial group average differences. (There has been some success using regression analysis to identify associations between environmental factors and differences between the mean test scores for racial groups; Fryer and Levitt 2004).

Many human behavioral geneticists and psychometricians (analysts of data from personality and educational tests) are prepared to entertain a two-part argument: the high heritability of IQ test scores within racial groups *coupled with* a failure of environmental hypotheses to account for the group differences supports explanations of mean differences in terms of genetic factors (e.g., Jensen in Miele 2002, 111ff). (The specific factors still have to be elucidated, so "support" may be read as "lends plausibility to the belief that they exist.") By that logic, however, we would have to entertain explanations of *generational* differences in terms of genetic factors, but we know that the changes in gene frequencies in a human population over one generation are negligible. There must be a hole in the logic of the two-part argument, but where is it? If we were to find the hole, would that help us explain large differences between generations in a high heritability trait such as IQ test score? These questions constitute the *IQ paradox* of Dickens and Flynn (2001).

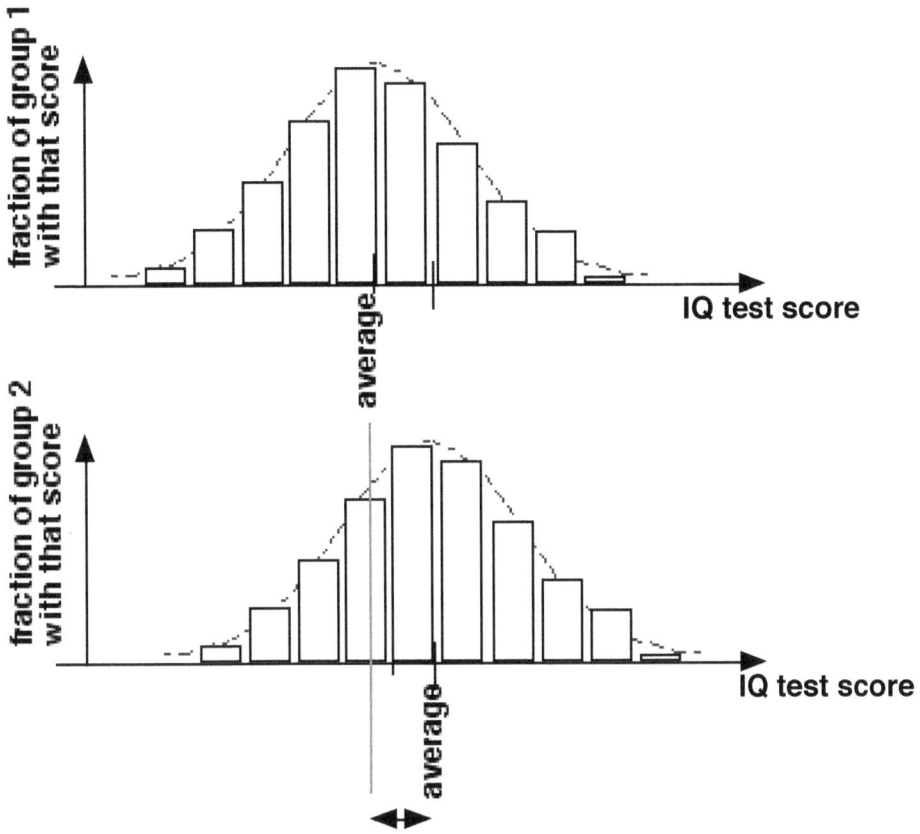

Figure A.1. Schematic of the "Flynn effect"—large gains in average IQ test score between generations, marked group 1 and group 2 in this figure.

Puzzle 2. The possibility of underlying heterogeneity

Claims that some human trait, say, IQ test score at age 18, shows high heritability derive from an analysis of data from relatives. For example, the similarity of pairs of monozygotic twins (who share all their genes) can be compared with the similarity of pairs of dizygotic twins (who do not share all their genes). The more that the former similarity exceeds the latter, the higher the trait's *heritability*. Researchers and commentators often describe such calculations as showing how much a trait is *heritable* or *genetic*. However, no genes or measurable genetic factors (such as alleles, tandem repeats, or chromosomal inversions) are examined in deriving heritability estimates, nor does the method of analysis suggest where to look for them. Moreover, even if the similarity between twins or a set of close relatives is associated with the similarity of yet-to-be-identified genetic factors, *the factors may not be the same from one set of relatives to the next, or from one environment to the next.* In other words, the underlying factors may be *heterogeneous.* It could be that pairs of alleles, say, AAbbcbDDee, subject to a sequence of environmental factors, say, FghiJ, are associated, all other things being equal, with the same outcomes as alleles aabbCCDDEE subject to a sequence of environmental factors FgHiJ (see Figure A.2 for the case of human twins where both members of each pair are raised in the same household). If the genetic and environmental factors underlying the observed trait are heterogeneous, what can researchers do on the basis of a heritability estimate? What can researchers they do when the underlying factors are unknown and the method of data analysis does not rule out the possibility that those factors are heterogeneous?

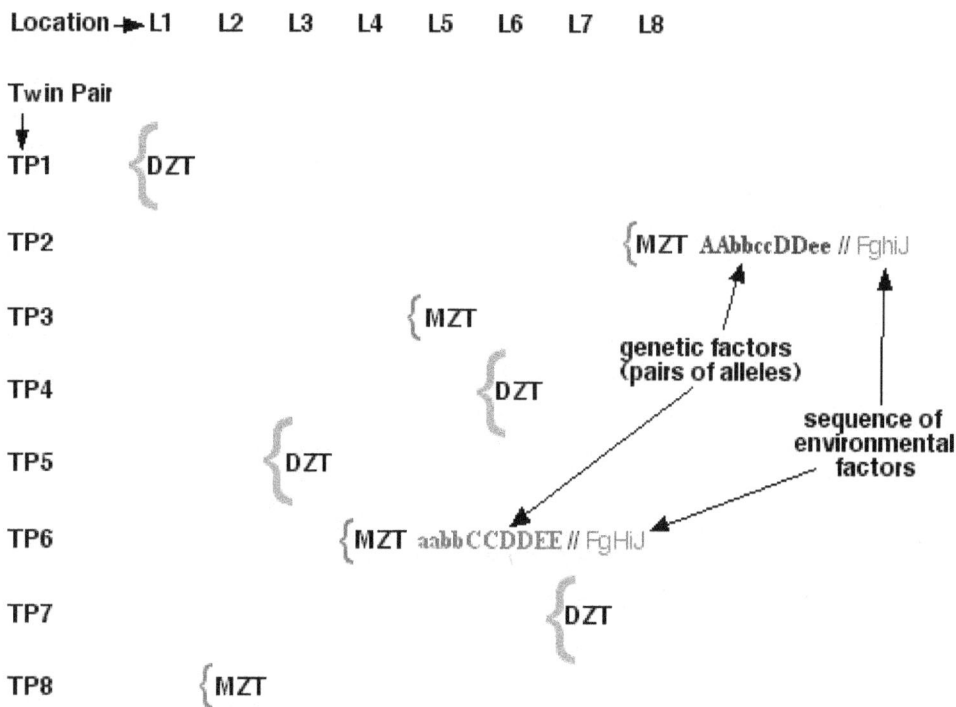

Figure A.2. Factors underlying a trait may be heterogeneous even when identical (monozygotic) twins raised together (MZT) are more similar than fraternal (dizygotic) twins raised together (DZT). The greater similarity is indicated here by smaller size of the curly brackets. The underlying factors for two MZT pairs are indicated by upper- and lowercase letters for pairs of alleles (A-E) and the environmental factors to which they are subject (F-J).

Puzzle 3. The potential for confusion in translating from finding patterns in data to selective breeding and the manipulation of underlying factors

Many of the terms and concepts in the statistical analysis of trait variation arose in agricultural trials directed toward selective breeding or toward the manipulation of environmental factors that influence the development of traits of interest. In other words, *patterns* in data are translated into ideas about the *factors influencing the dynamics* of development and reproduction under selective mating. These translations create the potential for confusion.

Heritability connotes a connection between parent and offspring through transmission of genes. However, as mentioned in Puzzle 2, the term's technical meaning and statistical estimation involve no reference to measurable, transmissible genetic factors. The estimation does not even entail a connection between one generation and the next—heritability for a given trait is defined as a quantity that summarizes observations made of a specific set of varieties and locations *at one point in time*. (Heritability can be calculated from correlations between parents and offspring, but to do so requires models of theoretical, unobserved genes that determine the trait, and a suite of other assumptions; see Gaps 3-6 in Item B and Item E.)

If heritability invites confusion, so does the related term *genetic variation* or variance (where *variance* is the technical way of quantifying variation). This term might seem to refer to variation among different individuals who possess some measurable genetic factor(s). However, the so-called genetic variation that is involved in the estimation of heritability is the variation across groups of related individuals (i.e, across *varieties* or *genotypes*) in the average value of the *trait* for each group. This distinction has been muddied by the increasing use over the last few years of the term heritability to refer to a conceptually and empirically distinct quantity that summarizes variation in measurable genetic factors. This new heritability is the fraction of variation in a trait not associated with variation in Single-Nucleotide Polymorphisms (SNPs) as examined by Genome-Wide Association studies (see Items H.6 and I.4).

The potential for confusion becomes even greater when researchers who are technically proficient in the statistical analysis of *measurements on a trait* interpret results of heritability studies in terms of the influence of the (unknown) measurable *genetic and environmental factors* that underlie the development of the trait. To cite just a few examples: Heritability is defined by

Layzer (1974, 1259) as the "fraction of the variance of a phenotypic trait in a given population caused by (or attributable to) genetic differences." As noted in the Introduction, Plomin et al. (1997, 83) define heritability as the "contribution of genetic differences to observed differences among individuals." Turkheimer (2000, 160), summarizing the findings of quantitative genetics about human behavioral or psychological traits, states: "The effect of being raised in the same family is smaller than the effect of genes."

How has the potential for confusion around heritability and genetic variation played out over the long history of the application of statistical analyses of trait variation in agricultural trials and selective breeding? (Recall that this is the context in which the methods of analysis in heritability studies arose.) When agricultural and laboratory breeders identify a *pattern* in measurements of a *specific set of individuals* in a *specific range of situations* at *one point in* time, in what ways do they use that pattern to identify *factors influencing the dynamics* of development and reproduction under selective mating? (Indeed, one might puzzle over the assumptions and steps needed more generally to connect pattern to process.) How much do the breeders' methods for translating patterns in data into measurable factors depend on the agricultural or laboratory context in which varieties and locations (often labeled genotypes and environments) can be deliberately chosen and replicated? What does this say about the interpretation of statistical analyses of human variation, where control and replication are difficult?

Item B. Eight gaps in analyzing and interpreting heritability

In which eight conceptual and methodological gaps, which researchers and commentators have not acknowledged or not appropriately responded to, are distilled from heritability studies.

To gain insight into the three puzzles in Item A, we should acknowledge and make appropriate responses to the eight conceptual and methodological gaps, listed in Table Intro.1, now introduced in sequence.

Gap 1. Key terms have multiple meanings that are distinct

Key terms have multiple meanings that are distinct and should not be conflated; in other words, this first gap needs to be kept open. Terminological changes can reduce the potential for confusion.

Consider, especially, the term *genetic*. Two different meanings of the term were identified in Puzzle 3 (Item A). To these we can add the common use of the term to mean *inherited* or to mean something that recurs in a family. To reduce the potential for confusion, the term *genetic* will be used here as an adjective and will refer to entities or factors that are transmitted through the germline from parents to offspring and whose presence or absence can, in principle, be observed. Similarly, *environmental* is used exclusively to refer to measurable factors.

Additional terminological steps that can be taken to reduce potential confusion include avoiding the term *phenotype* to refer to traits and *genotype* and *environment* to refer to groups of identical or related individuals and the locations or situations in which they are raised or grown. Those terms are prone to obscure the fact that analysis of variation in traits neither *requires* knowledge nor, on its own, *produces* knowledge about the genetic or environmental factors that influence the development of the individual's trait (phenotype) in the circumstances in which it is raised (see Gap 2). The agricultural terms *variety* and *location* are suitably neutral replacements for the terms *genotype* and *environment*. A variety is a group of individuals whose relatedness by genealogy can be characterized, such as the offspring of a given pair of parents, or a group of individuals whose mix of genetic factors can be replicated, as in an open-pollinated plant variety or pure (genetically identical) lines. A location is the situation or place in which the variety is raised, such as a family (for humans) or a specific experimental research station (for agricultural varieties). (Locations in agricultural and laboratory research are sometimes defined by the level of a specific environmental factor, such as the temperature at which a fruit fly is raised or the manure applied to fertilize potatoes, but the term is used in this book without assuming that is necessarily the case.)

Note that the classical *quantitative genetics* that is the focus in Parts I and II is not the analysis of genes or genetic factors, but the analysis of continuous variation in *traits* of humans, other animals, or plants in ways that take account of the genealogical relatedness of the varieties whose traits are observed. (See Items H and I for a discussion of quantitative methods that have begun to employ information about genetic factors.) However, for want of a recognizable alternative, the term quantitative genetics will be preserved in this book, as will *heritability*. (As noted in the Introduction, as a non-technical synonym for aspects of quantitative genetics, *heritability studies* is also used.) We should, however, stay mindful that using these terms risks perpetuating misleading connotations (see Puzzles 2 and 3). The potential for confusion is reduced, but not eliminated, by acknowledging the next gap.

Gap 2. Statistical patterns are distinct from measurable underlying factors

Estimates of heritability derive from statistical analysis of variation in <u>traits</u> among related and unrelated individuals. They do not reference measurable genetic or environmental <u>factors</u> involved in the development of those traits. It follows that any estimates or patterns detected by the analysis, such as the size of heritability relative to other fractions of the overall variation, must also be distinct from those factors. The gap between statistical patterns and measurable underlying factors needs to be recognized and, as with Gap 1, kept open.

The point is not new, but even critical commentators often fail to maintain the distinction (see Gap 1 and Item H). To demonstrate the distinction and, at the same time, make the meaning of heritability clear, consider the simplest case, which is an agricultural evaluation trial of a set of varieties raised in each of set of locations, and where there are two or more replicates in each variety-location combination (Figure B.1). The total variation among replicates for any

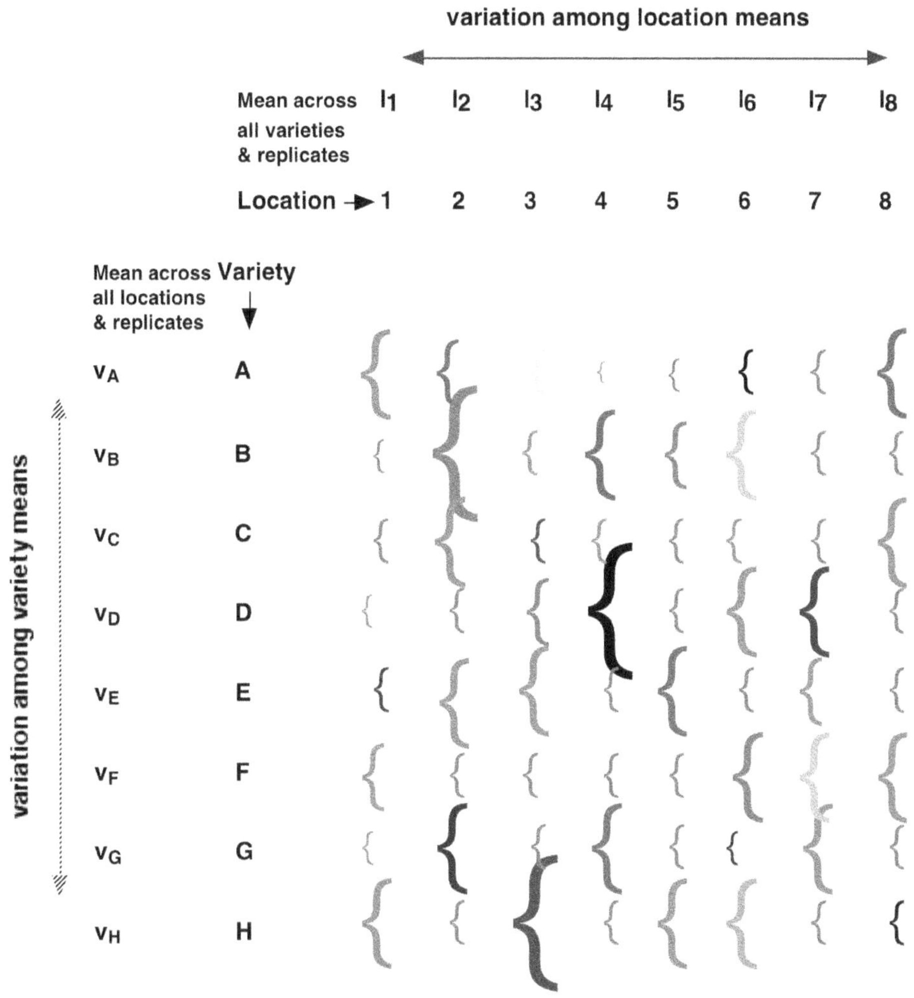

Figure B.1. Partitioning of variation for a given trait in the ideal agricultural evaluation trial where each of a set of varieties is raised in each of a set of locations, and there are two or more replicates in each variety-location combination. The variation between replicates within variety-location combinations is indicated by the size of the curly brackets. The non-systematic shading of the brackets indicates that the variation between replicates is not correlated from one variety-location combination to another. (The agricultural evaluation trial contrasts with Figure A.2 in which the replicates of any one variety—twin pairs—are raised in only one location—household—per variety.)

observed trait, such as crop yield per unit area, can be divided into four components: the variation among the variety means (v_A, v_B, etc.), the variation among the location means (l_1, l_2, etc.), the variation among means for variety-location combinations after subtracting the variety and location means for each combination (not indicated in Figure B.1), and the variation among replicates within variety-location combinations (indicated by the curly brackets). (This partitioning of variation is called the Analysis of Variance or ANOVA.) Heritability for the trait is, by definition, simply the first of the four components—variation among the variety means—expressed as a fraction of the total variation (see Box B.1 for elaboration).

In the spirit of spelling out the various connections between concept, method, and application related to each gap, let us review several questionable ideas and practices that contribute to obscuring the distinction between statistical patterns and measurable underlying factors.

a. The components of variation are given *shorthand names*—variety variance, location variance, variety-location interaction variance, and error variance. (Variance, again, quantifies variation. The variety variance, for example, assesses the size of the deviations of the variety means from the overall mean for the trait by averaging the square of those deviations.) Heritability becomes the ratio of the variety (or "genetic") variance to the total variance, which can be re-expressed *ambiguously* as the "proportion… that is attributable to genetic variation among individuals" (Wikipedia n.d., a). That formulation can lead to statements that obscure the distinction altogether, such as those noted in the Introduction and Puzzle 3. Here is another example: "Heritability analyses estimate the relative contributions of differences in genetic and non-genetic factors to the total phenotypic [i.e., trait] variance in a population" (Wikipedia n.d., a). Similarly, the distinction is obscured when location (or "shared environmental") variance gets interpreted as a measure of the effect of experiences or environmental factors shared by replicates in a variety-location combination (e.g., the members of a family growing up together; see Item H.3).

b. When estimating heritability from datasets in which varieties have varying degrees of genealogical relatedness (e.g., identical or monozygotic twins versus fraternal or dizygotic twins), models often refer to *theoretical genes* that each add a small contribution to the trait. However, analyses built around these models are of observations of *traits*, so there must always be alternative form-ulations making no reference to genes (Items D and E; see also Gaps 4 and 5).

Box B.1. Heritability defined and estimated

Heritability is the variance among the variety means expressed as a fraction of the total variance for a trait as observed for individuals raised in a given set of variety-location combinations. Strictly, this defines *across-location* heritability. An alternative, *within-location* heritability, is relevant where researchers expect that the variety will continue to be raised in the same location. In effect, this quantity takes the heritability estimated in *each location separately* and averages these estimates over all locations. This method of estimation does not take into account differences between the averages for the trait from one location to the next. Across-location heritability, which always has a smaller value than within-location heritability, is relevant when the varieties could be raised or grown again in any of the original set of locations.

Strictly, variation among the variety means defines heritability in the *broad* sense. *Narrow-sense* heritability, which is used to predict change under selective breeding, is defined in a way that depends on assumptions about the action of theoretical idealized genes in the standard models of quantitative genetics (see Item E.1 and E.4.1.b).

Heritability can also be estimated through path analysis, a data analysis technique that quantifies the relative contributions (*path coefficients*) of variables to the variation in a focal variable once a certain network of interrelated variables has been accepted (Lynch & Walsh 1998, 823). (*Variable* here may refer to measurable factors or to quantities [so-called *effects*] defined by an Analysis of Variance [see Item F.5].) The reliability of the estimates depends on the assumptions built into the networks, such as the similarity of relatives of different degrees and the inclusion or exclusion of coefficients for variety-location interaction (see Gap 5 and Item F.5). When the Analysis of Variance and path analysis use the same assumptions, they estimate the same quantities, including heritability.

Recently, heritability has been used to refer to the fraction of variation for a trait accounted for by genomic variants identified in Genome Wide Association studies (Manolio et al. 2009). When this quantity is mentioned in this book, the label *new heritability* will be used to mark that it is unrelated to heritability as defined above (Item H.6).

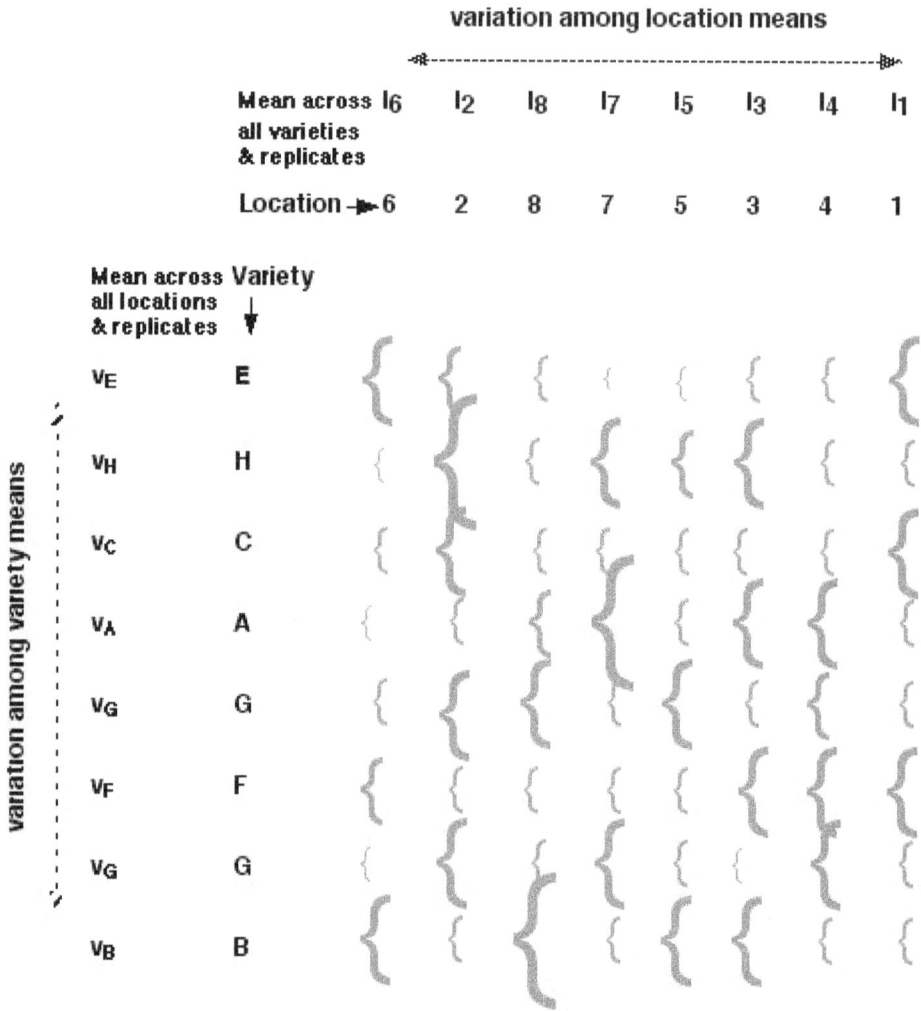

Figure B.2. Partitioning of variation related to hypothetical genetic and environmental gradients. The arbitrary order of varieties and locations in Figure B.1 has been replaced by a specific order along these gradients given by the relative size of, respectively, the variety means and the location means. See text for an argument against assuming such gradients exist.

c. A *gradient* of a measurable genetic factor (or composite of factors) is assumed to run through the differences among variety means (Figure B.2 and Box B.2). To see that this assumption is not necessary, note that the analyses in heritability studies do not require the varieties to be from the same species or even the same taxonomic class. If the varieties were not, we would not assume such a genetic gradient exists. Yet, there is nothing in the method of data analysis that changes from a multi-species situation to one in which the varieties

Box B.2 Notation and hypothetical gradients

The existence of gradients in genetic and environmental factors underlying the variation in traits is suggested by the symbols often used in equations, e.g., $P = G + E$, where P is used to denote the measurement of the trait (or phenotype), G a contribution from the variety (or genotype), and E is a contribution from the location (or environment or environment plus error). Such contributions are estimated, however, using statistical methods, such as ANOVA and path analysis, that partition the variation in *traits* across some specific set of varieties and locations into components *without* reference to data about underlying genetic and environmental gradients.

are drawn from a single species or even to one in which varieties are inbred lines. There is no point in such a shift at which we gain license to assume that an underlying gradient is present. Similarly, except in agricultural and laboratory research where a location is defined in terms of a specific environmental factor, there are no grounds to assume that a gradient of environmental factors runs through differences among location means.

d. The *conditionality* of patterns derived from statistical analyses of traits is discounted. Conditionality means that, with a different set of varieties and locations in the data, a different partitioning of variation results. Granted, a variety mean does serve as the single value of the trait for each variety that best conveys its average difference from other varieties (see Figure B.1). This might appear to make the assumption in c. plausible. However, the variety mean is an average calculated over a particular range of locations in which varieties are observed, so the variety mean is not simply a property of the variety. Conditionality likewise applies to location means.

e. Heritability is given *causal significance* without identifying the measurable genetic and environmental factors that underlie it. In the broad sense of causes as differences that make a difference, a difference among the observed trait for individuals from two varieties can be associated with a difference between two variety means. Heritability has causal significance in that sense given that it quantifies the average of the variety-mean differences (squaring them so the direction of the difference becomes unimportant). Similarly, differences between location means can be thought of as causes. However, the insight that can be drawn from assessing how much traits are caused in this sense is limited. Because the variety and location means are conditional on the particular sets of varieties and locations, this difference-between-means form of causality corresponds to a situation in which the only thing that can vary from the original to any *rerun* is the *noise* (i.e., unsystematic, residual, or error variation in an ANOVA). (This point is developed in Item F).

f. *Replicability* of varieties and locations is imagined for human traits (see Item F.4 for discussion).

g. The idea of *genotype-environment correlation* (or gene-environment correlation), which has a well-defined technical meaning in quantitative genetics (Jacquard 1983; Lynch and Walsh 1998, 47), is used in more colloquial discussions that assume that genes can elicit particular environmental factors or vice versa (Plomin et al. 1977, Scarr and McCartney 1983, Sesardic 2005) (see Item H.2 for discussion).

If we do *not* adopt the preceding ideas and practices, but instead highlight the gap between statistical patterns in traits and underlying measurable factors, we might ask how the gap can be bridged. This question leads to the third gap.

Gap 3. Translation from statistical analyses to hypotheses about measurable factors is difficult

It is difficult to translate from statistical analyses of data on traits to hypotheses about the measurable genetic or environmental factors that influence the development of the traits. The steps and conditions needed to bridge this gap—or to circumvent it—warrant attention.

Consider three paths that researchers might pursue:

a. Undertake research to identify the specific, measurable genetic and environmental factors **without reference to the trait's heritability** or the other fractions of the total variance (e.g., Moffitt et al. 2005, Davey Smith and Ebrahim 2007, Khoury et al. 2007). Discussion of this direction of research takes us beyond heritability studies (see Item I).

b. Use high heritability as an indicator that "the trait [is] a potentially worthwhile **candidate for molecular research**" that would identify the specific genetic factors involved (Nuffield Council on Bioethics 2002, chap. 11; emphasis mine). This path assumes that a gradient of a measurable genetic factor (or composite of factors) runs through the differences among variety means (*contra* Gap 2, point c., above). There may well be certain traits for which such a gradient exists. These traits might be worth finding even if, in the course of doing so, researchers end up conducting fruitless investigations of other heritable traits for which it turns out there is no such gradient. (This search for the traits with a genetic gradient is not a search for traits that are largely determined by genes at a single locus, whose effect is more or less independent of the individuals' upbringing. [Presence of extra digits or polydactyly would be such a trait.] Such *high penetrance major genes* can be detected through examination of family trees; heritability analysis need not be involved.) In pursuing molecular research guided by heritability estimates, there is a risk that the proportion of fruitful investigations will be low compared to investigations confounded by the lack of an underlying gradient. In any case, additional knowledge beyond that derived from the statistical analysis is always required.

    c. Restrict attention to variation ***within a set of relatives***. Even if the underlying factors are not yet known, high heritability still means that if one twin develops a trait (e.g., type 1 diabetes), the other twin is more likely to as well. This information might stimulate the second twin to take measures to reduce the health impact if and when the disease starts to appear. However, notice that this path assumes that the timing of getting the condition differs from the first twin to the second. Researchers might well then ask: What factors influence the timing? How changeable are these? How much reduction in risk comes from changing them? To address these issues researchers would have to identify the genetic and environmental factors that influence the development of the trait. To do so would require larger sample sizes than any single set of relatives allows. The question then arises whether the initial results would carry over from one set of relatives to others. This issue is an empirical one; there is a risk, as before, that the proportion of fruitful investigations will be low compared to those confounded by factors not carrying over well from the initial set of relatives.

    It should be noted that path b. seems to rest on an intuition that high heritability indicates that measurable genetic factors have more influence on variation in the trait than measurable environmental factors (even though the specific factors are unknown); similarly for the ratio of variation among location means (or shared environmental effects) and other fractions of variance (Turkheimer 2000). However, this intuition is unreliable. It lacks support, even in the ideal case of agricultural evaluation trials. To gather such support would require, in the absence of prior knowledge of how genetic and environmental factors influence the development of the trait, a program of theoretical exploration consisting of several steps (which have not, to my knowledge, been undertaken): consider a range of models of factors influencing development; justify the assumptions on which the models are built; calculate heritability for a representative range of values of each model's parameters; and discover associations between the heritability calculations and the corresponding genetic and environmental factors that are robust across models (Taylor 2006a).

Now, to speak of considering a range of models is to imply that alternatives exist to the standard models presented in quantitative genetic texts, models that center on theoretical genes. Items D and E elaborate on the standard models and alternatives, including models free of references to theoretical genetic and environmental factors. For now, let us note that employing models of theoretical genes and giving advice to relatives are examples of trying to circumvent Gap 3. There are alternative actions that can be taken in the absence of knowledge of underlying factors; these are discussed in the next gap.

Gap 4. Predictions based on extrapolations from existing patterns of variation may not match outcomes

It is possible, even without knowledge of the underlying measurable factors that are reproductively transmitted and influence the development of the trait, to extrapolate from existing patterns of variation (as captured, for example, by heritability estimates). However, we can expect that this extrapolation imperfectly predicts the actual outcomes. If any actions depend on such predictions, we need to be able to compensate for the discrepancies between predictions and outcomes.

If we do not have knowledge of the measurable underlying factors we can focus on heritability as a fraction of the variation among measurements. This approach is evident in agricultural and laboratory breeding, where selection of parents of the next generation can proceed on the basis of observed traits and without knowledge of the underlying factors. High (or low) heritability is used to indicate whether (or not) to expect selective breeding to produce the desired improvement in the average value of the trait across the population. (Selective breeding is, of course, not an acceptable option for humans.) Moreover, simple models of multiple, theoretical genes underlying the traits (Item E.1) allow breeders to make more refined predictions of the advance under different breeding plans (e.g., mating of half-siblings versus mating of cousins), which can inform breeders' decisions about which plan to implement.

Because heritability is a summary of observations made at one point in time for a specified set of varieties and locations (Gap 2, point d.), we would expect it to be an imperfect predictor of advances from one generation to the next under selection (which changes the mix of varieties) and under breeding (which produces new genetic combinations). However, in agricultural and laboratory settings, researchers are able to replicate varieties and locations (see Gap 2, point f.) and they can select among varieties for the next generation on the assumption that the environmental factors will remain more or less unchanged. Because the researchers have such control, the predictions can be made with more confidence. In any case, if the actual advance under selective breeding is less than predicted, breeders can always compensate for discrepancies—they discard the undesired offspring, breed the desired ones, and continue.

Control of situations is, like selective breeding, neither acceptable nor achievable for humans. Nevertheless, when models of genes (albeit theoretical ones) are fitted to the observed variation in the trait, the models appear to show the relative degree of influence of genetic and other factors on the trait, even if the identity of those factors remains unknown. (In other words, Gaps 1 and 2—Terminology and Patterns versus factors—seem not to matter, and the intuition behind the two directions under Gap 3—Translation to hypotheses—seems to be justified.) By extension, high heritability leads us to expect only small changes following shifts in environmental factors. This last idea would seem to apply to humans as well, which would mean that implications could be drawn from patterns in variation even outside the realm of selective breeding.

The problem with expectations based on models of theoretical genes is that, in the absence of evidence for the model's assumptions and without comparing a range of alternative models (see discussion in Gap 3), the reliability of the predictions is uncertain, as are any claims about the relative degree of influence. This means, in short, that it is not so easy to sidestep the difficulty of translating from statistical analyses of data on traits to hypotheses about the measurable genetic or environmental factors that influence the development of the traits (Gap 3).

The situation is further limited, especially for human studies, by the next gap.

Gap 5. The partitioning of variation, especially in human studies, does not reliably estimate the intended quantities

The conventional estimation of heritability (and of other fractions of the trait's variation) is not reliable because it depends on certain fundamental assumptions. The results and interpretation can change profoundly if plausible alternative assumptions are chosen. To remedy the gap between the actual estimation and reliable results requires data sets that are difficult to collect.

The fundamental assumptions are listed in Box B.3 and elaborated in Item E. It should be noted that this gap goes beyond Gaps 3 and 4—Translation to hypotheses and Predictions—which imply that for humans, where selective breeding and control of situations are not acceptable or achievable, high heritability for a trait is relevant only to restrict attention to variation within a set of relatives (path c. in Gap 3) or as a possible indicator that the trait is a candidate for molecular research, where its relevance rests on an intuition that is difficult to justify about the relative degree of influence by genetic and environmental factors (path b. in Gap 3).

Box B.3 Assumptions in standard quantitative genetic analysis

These assumptions include the following:

a. The analysis requires models of genes with simple Mendelian inheritance and direct contributions to the trait.

b. All other things being equal, similarity in *traits* for relatives is proportional to the fraction shared by the relatives of all the *genes* that vary in the population. (In studies of human twins, fraternal or dizygotic twins are assumed to be half as similar, all other things being equal, as identical or monozygotic twins.)

c. Variance among variety-location-combination means (the so-called genotype-environment-interaction variance) can be discounted in human studies or can be incorporated into the heritability estimates.

d. Residual variance in human studies is a within-family *environmental* contribution (the so-called non-shared environmental effects).

e. When similarity among close relatives (such as twin pairs) is associated with similarity of (yet-to-be-identified and measured) genes or genetic factors, those factors are the same from one set of relatives to the next.

If we do not make the last assumption in Box B.3 (which had been introduced in Puzzle 2), we are led to the next gap.

Gap 6. Translation from statistical analyses to hypotheses about the measurable factors is even more difficult in light of the possible heterogeneity of underlying genetic or environmental factors.

Even when similarity within a set of close relatives (such as twin pairs) is associated with similarity of (yet-to-be-identified and measured) genetic or environmental factors underlying the development of the trait, it is possible that those factors are not the same from one set of relatives to the next. If the genetic and environmental factors underlying the observed trait could be heterogeneous, it is even more difficult to translate from statistical analyses of data on traits to hypotheses about the measurable genetic or environmental factors that influence the development of the traits. The steps and conditions needed to bridge this gap—or to circumvent it—warrant attention.

(This gap was noted in the Introduction and in Puzzle 2 of Item A; it runs counter to assumption e in Box B.3 above. On visualizing underlying heterogeneity, see Figure A.2 and Item G.0. For further discussion, see Items G, I, and J.)

Researchers might try to bridge or circumvent the knowledge that underlying factors may be heterogeneous by choosing one of the following six paths. (Difficulties in translation to hypotheses and in making predictions, over and above those discussed in Gaps 3 and 4—and more serious than those difficulties—are evident in these paths.) The first four paths parallel those in Gaps 3 and 4:

a. Undertake ***research*** to identify the specific, measurable genetic and environmental factors ***without reference to the trait's heritability*** or the other fractions of the total variance. Discussion of this research direction takes us beyond heritability studies, as was noted already in Gap 3. Yet, as will be discussed in Part III, how to pay attention to the possibility of heterogeneity is a challenge for research on heredity and for the human sciences more generally.

b. Use high heritability (perhaps corrected in light of Gap 5—Unreliable estimates) to ***guide molecular research*** to identify the specific genetic factors involved. There may be traits for which the underlying factors are *not* heterogeneous. These traits and factors might be worth finding even if researchers do not know in advance what proportion of investigations will be fruitful and what proportion will be confounded by the underlying heterogeneity. Again, the search is not for high-penetrance major genes. Researchers would be looking for traits in which many underlying genetic factors, each of small influence, turned out to be similar for all individuals within some defined population who show the same value for the trait.

c. Restrict attention to *within a set of relatives*. The same logic described in Gap 3, path c reveals that the differential timing of getting the condition becomes an issue. Again, researchers have to identify the genetic and environmental factors that influence the development of the trait and to employ larger sample sizes than any single set of relatives. The question of what to do about the possibility of underlying heterogeneity of these factors thus persists.

d. Put aside the search for measurable factors. Instead, focus on *heritability as a fraction of the variation* among measurements, a focus that makes sense in agricultural and laboratory breeding. If the actual advance under selective breeding is less than predicted, one source of the discrepancy might be the underlying heterogeneity of genetic factors and their reassortment through mating. Again, this matters little because breeders can always compensate for discrepancies: they discard the undesired offspring, breed the desired ones, and continue. Selective breeding is not an acceptable option for humans, so heritability as a fraction of variation is only useful under the interpretation that genetic factors have a larger influence than environmental factors for high heritability traits. As discussed in Gap 3, support for such an interpretation is lacking; the theoretical exploration required would be all the more challenging if we considered models that allow for heterogeneous factors that underlie the trait.

Researchers can also address the possibility of underlying heterogeneity in two ways that are not discussed under Gaps 3 and 4:

e. Reduce the possibility of underlying heterogeneity by *restricting the range of varieties or locations*. Agricultural researchers can restrict the range of locations in which a variety is raised or grown. They can also control environmental conditions, such as (for animals) the regimes of feeding and husbandry or (for plants) the application of fertilizer and irrigated water. Agricultural breeders can often produce inbred lines and thereby eliminate the heterogeneity of genetic factors that characterize outbred varieties. However, to envision taking action on the basis of research conducted under restrictive conditions is to presume that the restrictive conditions can be replicated. This presumption is most apparent when plant breeders recommend varieties that are to be grown only in defined regions and under prescribed techniques of cultivation, or when animal breeders specify the optimal feeding and husbandry for each variety. In the study of human traits, however, it is not feasible to control the full range of relevant environmental conditions or to breed for genetic uniformity. There may, nevertheless, be some ways to restrict the locations included in a human study (e.g., to include only families of low socioeconomic status; Turkheimer et al. 2003). The heritability estimates would be reliable (again perhaps after correction; see Gap 5) to the extent that these restrictions were replicated in subsequent research or policy. With restrictions replicated, the research could be applied even though the environmental factors underlying those locations had not been identified.

f. Reduce the possibility of underlying heterogeneity by *grouping varieties* that have similar responses across locations (see Item G.2 for discussion). This is not a feasible direction by which research on human variation can bridge or circumvent Gap 6. After all, when analyzing measurements from human twins, there are only two replicates (twins) in one or at most two locations (families).

Gap 7. Many steps lie between the analysis of observed traits and the creation of interventions based on well-founded claims about the causal influence of genetic or environmental factors

At each of the steps that make up this gap, conditions and assumptions apply that render the value of estimates of heritability and other fractions of the variation even more limited than has been discussed thus far.

Suppose that hypotheses have been derived about measurable factors (even though, as Gaps 3–6 suggest is the case for human traits, statistical analyses such as ANOVA provide little guidance). The next step for researchers is to investigate associations of the trait with measurable factors, which can mean either using regression analysis and related statistical techniques or conducting experimental trials. In both cases, conditionality (Gap 2, point d.) applies, now also extended to conditionality on the set of factors measured. Moreover, when researchers choose significant factors from a statistical analysis to be manipulated in experimental trials, they are assuming that this manipulation does not modify the structure of the overall dynamics within which the factors had been associated with the observed traits. (Manipulations or interventions that preserve the same dynamics seem more plausible for agricultural and laboratory trials than for human social relations; see Freedman 2005 and further discussion in Item I.3). Insights from these experimental studies can, in turn, contribute to research on the ways that pathways of growth and development are influenced by the genetic makeup of varieties and the environmental factors in the locations. Such research might, in turn, provide a basis for interventions outside the typically well-controlled conditions in which research on causes in growth and development is undertaken. The sequence of steps in this paragraph is summarized in Table B.1.

Gap 8. Explanation of variation within groups does not translate to explanation of differences between groups

Accounting for within-group variation does not explain between-group differences. This is widely acknowledged, but, in the contentious debates about differences in the averages for racial and other groups, high heritability is seen by some to confer plausibility to hypotheses about the role of genetic factors in explaining those differences (see Puzzle 1). The gap between within-group and between-group differences is, however, firm and its implications are deep.

The following considerations counter the plausibility of hypotheses that invoke genetic factors to explain between-group differences (Jensen in Miele 2002, 111ff, Sesardic 2005):

a. Statistical analysis of variation among traits, which includes heritability estimation, provides little or no guidance in hypothesizing about measurable factors behind observations of human traits within one group of varieties (Gaps 3-6), so it can provide little or no guidance about measurable factors associated with differences between two groups. This point alone discounts the relevance of heritability to discussions of group differences (Taylor 2006b). Referring back to Puzzle 1, the two-part argument about IQ test scores changes form and loses its bite: There may be no measurable environmental factor associated strongly with the group or generational average differences, *but there is no such genetic factor either*. (Associations have not been found in the few instances where genetic factors have been examined, e.g., genes that mark degree of African ancestry; see summary in Nisbett 1998, 89–90.) The average group or

Table B.1. Connections from one kind of data analysis to the next

Kind of data to be analyzed	Agricultural evaluation trials (varieties each replicated over a number of locations)	Human studies of twins and other relatives
Observations of a trait that differs across different varieties and locations ↓	ANOVA + cluster analysis + knowledge from sources outside data ↓ Hypotheses about measurable factors	ANOVA (& path analysis) not helpful in generating hypotheses about measurable factors[a] (Hypotheses about factors drawn from other sources)
Observed associations with measurable factors ↓	Factors significant according to regression analysis ↓ Factors for testing through experimental trials	Factors significant according to regression analysis ↓ Factors for testing in experimental trials[b]
Experiments that vary measurable factors ↓	Significant factors ↓ Insights for investigating dynamics of development	(Rare) ?
Factors observed over the course of development (rarely-realized ideal)	Significant factors in development under controlled research conditions ↓ Candidates for interventions in less controlled situations	? ?

a. The solid line underneath denotes the disconnect between the data analysis and the generation of hypotheses about measurable factors.
b. The manipulation of factors without modifying the structure of dynamics is more questionable for humans than for agricultural species.

generational differences still need explanation, but high heritability—if it is truly high (see Gap 5—Unreliable estimates)—poses no paradox.

b. For agricultural evaluation trials in which the observations of the trait are used to cluster varieties by similar responses across locations (Gap 6, path f; discussed in Item G), the possibility of underlying heterogeneity is minimized. This in turn allows researchers to hypothesize about the group averages, that is, about what factors in the locations elicited basically the same response from varieties in a particular variety group, responses that distinguish one group from another. However, if varieties are grouped by some criterion other than similarity of responses across locations, the possibility that heterogeneity of underlying factors is present can no longer be discounted when we want to translate from patterns in the data to hypotheses (Gap 6). In this situation the relationship between factors and patterns in data that underlies any hypothesizing may be difficult to disentangle (discussed in Item G).

c. Lindman's (1992) textbook illustrates a cautionary note about *nested* ANOVA (i.e., when each variety is replicated in one location only), using high school students' test scores in algebra viewed in relation to their teacher and school as an example. The students within a school are randomly assigned to a teacher in that school. Lindman notes that a significant difference between the means for the locations (schools) "is likely to be interpreted as due to differences in physical facilities, administration, and other factors that are independent of the teaching abilities of the teachers themselves… [However, d]ifferences between teachers in different schools are part of the [average location or school difference], and the observed differences between schools could be due entirely to the fact that some schools have better teachers [or] some schools have smarter children attending them" (Lindman 1992, 194).

Lindman could have added that the observed differences between schools could be due to any combination of factors, such as students respond worse to teachers whose attention is distracted because their school's administrators insist more on detailed documentation of student performance. Nested ANOVA cannot help researchers hypothesize about the difference in the average scores from one school to the next when the teachers are replicated (in their students' test scores) *only within schools*. To apply this to the concerns of this gap, nested ANOVA cannot help researchers hypothesize about the difference in the average scores from one location (or subset of locations) to the next when the varieties are replicated only within locations (or subsets of locations). Researchers might just as well conduct a separate ANOVA for each subset of varieties and locations—or, in the context of racial differences, for each racial group and the experience that members have of being in that racial group. (To respect this *methodological* limitation of nested data analysis is not to make the claim that disjunct kinds of causes *must* be operating in the different racial groups.)

d. Another consideration relevant to the within-group versus between-group gap is the relationship between lack of attention to the possibility of heterogeneity of underlying factors (see Gap 6) and a *typological or essentialist worldview* that "conceptualizes diversity as 'deviation' from a natural state or path of change" (McLaughlin 1998, 25). Notice that Lindman, even as he performs the valuable role of cautioning readers about nested analyses, perpetuates the typological worldview as he writes about "the observed differences between schools" when referring to the observed differences between *averages for schools*. It is still commonplace to hear typological expressions of the kind "men are taller than women," "men tend to be taller than women," or "men are, on average, taller than women." Some might dispute the label typological, saying that the implicit variation is understood. They might see little to be gained by wordier statements that make the variation explicit, such as, "the variation among men's heights centers at a point that is greater than the center of the variation among women's heights," or "the variation among men's heights and the variation among women's heights overlap, but some of the men's variation lies to the right of the women's variation and some of the women's lies to the left of the men's." However, can we be sure that it is simply linguistic convenience to use simple expressions that put group or class membership first and deviation as implicit or secondary? If not, the wordier, non-typological alternatives help keep in view the possibility that the factors underlying the pattern in the data could vary among men and women and *need not include factors solely possessed by one sex or the other* (see Item G.0 for further discussion). The wordier alternatives are more likely to steer us away from thinking that there is something essential in each group that leads to differences in averages from one group to the next. The wordier alternatives might even lead us to ask just who is empowered to do something as a result of an analysis of differences in group averages (or who is given license not to have to do anything)? These larger sociological questions are among the questions raised when the next Item revisits the three puzzles of Item A.

Item C. Resolving some puzzles and raising others

In which the three puzzles in Item A are revisited, in reverse order, presenting each in a new light and pointing to several new puzzles that invite attention from analysts of variation in quantitative genetics and social science.

Puzzle 3 revisited: The potential for confusion in translating from finding patterns in data to selective breeding and the manipulation of underlying factors

The eight conceptual-methodological gaps laid out in Item B, taken together, mean that the classical methods of quantitative genetics provide very little that is reliable and useful regarding the genetic and environmental factors that underlie traits, especially human behaviors and other traits. To recapitulate: Even in agricultural and laboratory breeding, where varieties and locations (often called genotypes and environments) can be controlled and replicated, the translation from statistical analyses to hypotheses about measurable factors is difficult. The translation is easier to achieve when the range of varieties or locations is restricted or when varieties that are similar in responses across locations are grouped (Gap 6, paths e. and f; see also Item G.2), but even then, knowledge from sources other than the data analysis is needed to help researchers generate hypotheses. Hypotheses, in turn, are only one step toward interventions based on well-founded claims about the causal influence of genetic or environmental factors (see Gap 7).

The difficulty of exposing the genetic and environmental factors that underlie traits can be circumvented by agricultural and laboratory breeders. Suppose they use the standard quantitative genetic models, which refer to theoretical genes that each make a small contribution to the trait, to predict advances under selective breeding. Then, even if the assumptions behind the models are impossible to verify or are unreliable (see Gaps 3–5 and Item E), the breeders can compensate for discrepancies: discard the undesired offspring, breed the desired ones, and continue. This option is unavailable to researchers studying human variation.

In short, the most important insight from Puzzle 3 is that the methods of quantitative genetics do not translate well from agricultural and laboratory breeding to statistical analyses of human variation. This conclusion points to new puzzles for historians, sociologists, and philosophers of biology. Given that human heritability estimation is based on data that are less ideal than agricultural evaluation trials, how has it been envisioned that human estimates could support claims about more general notions of genetic or environmental causality? How were restrictive conditions—the control and replicability that can be achieved in agricultural and laboratory breeding—discounted or forgotten when methods of heritability estimation were adapted to human genetics? (see Item F.6)

Puzzle 2 revisited: The possibility of underlying heterogeneity

The recapitulation of Item B in the first two paragraphs above also speaks to Puzzle 2, which concerns what researchers can do without knowing if the genetic or environmental factors that underlie traits are heterogeneous. In agricultural and laboratory trials, researchers can pursue approaches that make that possibility less disturbing—they have the ability to replicate varieties and locations, to control the variability in those varieties and locations, to reduce heterogeneity through grouping varieties by similarity of responses across locations, and to compensate for shortcomings in predictions of advances under selective breeding. In human studies, however, high heritability may be used primarily as a guide to decide whether to pursue molecular research to identify the specific genetic factors involved for the trait (Gap 6, path a.) or whether *not* to search for environmental influences or promote social policies based on them. Yet, using heritability to guide molecular or social research is likely to be fruitful only if three questionable conditions apply: the heritability is truly high (but see Gap 5), a gradient of a measurable genetic factor (or composite of factors) runs through the differences among variety means (but see Gap 2, point c.), and the underlying factors are not heterogeneous (but see Gap 6).

When the preceding conditions cannot be assured, it would be prudent for researchers not to place too much stock in heritability as a guide, let alone rely on the misleading intuition about a genetic gradient that may underlie variation among variety means (see Gaps 2, point c. and Gap 3). Instead, researchers could explore methods that attempt to identify the specific, measurable genetic and environmental factors without reference to the trait's heritability or the other fractions of the total variance (i.e., path a. under Gaps 3 and 6; see also Item I). This said, genomic studies have had very limited success identifying causally relevant genetic variants behind variation in human traits (McCarthy et al. 2008, Couzin-Frankel 2010), a result that is not be surprising if there is heterogeneity in the factors underlying most traits. Some genomics researchers have responded to the difficulties by emphasizing *genetic* heterogeneity (McClellan and King 2010). However, the challenge that researchers face identifying causally relevant factors is even greater because underlying heterogeneity encompasses environmental as well as genetic factors.

If the answer to Puzzle 2 is that only in circumscribed situations can researchers do anything reliable when they do not know whether or not the genetic or environmental factors that underlie traits are heterogeneous, a number of follow-up issues arise. First, it would be interesting to revisit studies that interpret heritability and so-called genetic variance as measuring the contribution of the genetic factors to the process through which the trait develops. What can be learned from the data and analyses in such studies once we highlight the gap between, on the one hand, the statistical analysis of measurements on a trait for a specific set of individuals in a specific range of situations, and hypotheses about the underlying genetic and environmental factors (Gap 2), on the other? (See Part II for a relevant reworking of quantitative genetic models and causal claims.)

It would also be interesting to extend the concern about underlying heterogeneity (Gap 6) to human sciences more generally. What shortcomings of current methods of data analysis and interpretation might emerge if researchers questioned the methodological assumption that, when similar responses of different individual types are observed, similar conjunctions of genetic and environmental factors have been involved in producing those responses? Similarly, we might examine the heterogeneity underlying risk and protective factors in epidemiology (Items I and J).

A final follow-up to Puzzle 2 stems from observing that, although some prominent geneticists have noted that heritability estimates are not helpful in identifying specific genetic factors (e.g., Rutter 2002, 4), the possible heterogeneity of factors that underlie patterns in observed traits has not yet been recognized as a significant issue, not only by quantitative geneticists but also by philosophers and other critical commentators on heritability research. For example, it is not mentioned as an issue in the extensive entry on heredity and heritability in the *Stanford Online Encyclopedia of Philosophy* (Downes 2004), in the key sources cited therein (e.g., Sarkar 1998; Kaplan 2000), or in the rebuttal of many critiques of heritability studies by Sesardic (2005). What conceptual and sociological considerations have obscured the heterogeneity issue during decades of debate? Could it be that philosophers of science have obscured the relevance of heterogeneity through visual and verbal conventions that emphasize types over variation? Perhaps they have they relied too much on scientists to set the terms of issues on which they focus their efforts in conceptual reconstruction? Similarly, perhaps sociologists have stuck too closely to the issues as debated by the scientists and the critics at the expense of delving into, as Williams (1980, 70) put it, the effects of concepts that the "opposing intellectual armies" share? How then, we might inquire, have each of the eight gaps been addressed or overlooked in previous studies and critiques?

Puzzle I revisited: The two-part argument and the IQ paradox

Having revisited Puzzles 2 and 3, we can return to Puzzle 1, which concerns average differences across generations and groups. The full puzzle is not solved here, but attention to the gaps allows us to dissolve the IQ paradox. Recall the two-part argument: the strong role of genetic factors within a group, coupled with the failure of environmental factors to explain differences among the average IQ test score for racial groups, lends plausibility to the idea that genetic factors are needed in an explanation. This plausibility depends, however, on interpreting high heritability within groups as evidence that genetic factors are more significant than environmental factors. When statistical patterns are seen as distinct from measurable underlying factors (Gaps 1 and 2) and we acknowledge the difficulties in translating from the patterns to hypotheses about the factors (Gaps 3 and 6), this interpretation of heritability no longer holds. The two-part argument can then be put to the side, so there is no paradox. (This can be said even without invoking the unreliability of heritability estimates for humans [Gap 5]; the many steps between hypotheses and intervention based on well-founded claims [Gap 7]; and the within-group/between-group disconnect [Gap 8].)

Although the specific IQ paradox is dissolved, the large average differences between groups and between generations on IQ test scores remain to be explained. The puzzle becomes: How do we identify the mix of genetic and environmental factors associated with those differences? Dickens and Flynn (2001) propose the use of *reciprocal causation* models, which involve two key features (Figure C.1): a matching of environments to differences that may initially be small (e.g., children who show an earlier interest in reading will be more likely to be given books and receive encouragement for their reading and book learning); and a social multiplier through which society's average level for the attribute in question influences the environment of the individual (e.g., if people grow up and are educated with others who, on average, have higher IQ test scores, this will stimulate their own development).

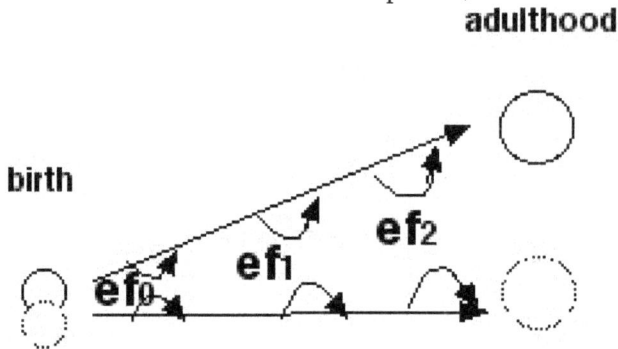

Figure C.1 Dickens and Flynn's (2001) Reciprocal Causation model (as depicted by the author). Small variation at birth is amplified by growing up in locations 0, 1, and 2 with environmental factors ef_0, ef_1, ef_2 that in part match the differences in the trait and in part result from transient non-matching influences. In addition, the environmental factors in every individual's location follows society-wide trends that result from the average of the changes across all individuals. The ranking of individuals at adulthood (or whenever the trait is measured) is correlated with the ranking at birth, but generation-to-generation trends can occur.

However, once it is recognized that the potency of social multipliers depends on the different capacities of various groups to capitalize on historical changes in society, there is no reason to assume that the multipliers apply uniformly across individuals regardless of their differences in age, gender, geographical location, culture, and so on, or even that the multipliers move different individuals in the same direction albeit at different speeds. Adapting a basketball analogy that Dickens and Flynn use to illustrate their reciprocal causation model: The onset of TV coverage of basketball acted as a social multiplier by eliciting greater participation in basketball; at the same time, it elicited more couch-potato spectators. In more general terms, if researchers envision developmental pathways made up of components that might be quite different from one group of individuals to others, the challenge is to develop methods to collect and analyze the data so as to discriminate among possible reciprocal causation and other models (see Item J).

The possibility that heterogeneous pathways underlie the variation in any given human trait leads, in turn, to a puzzle for socially engaged researchers (Taylor 2006b; Flynn 2000). If genetic factors are to be included in the models of trait development, there are good methodological reasons for *not* categorizing individuals according to racial group membership (e.g., this grouping is not based on clustering across a range of locations [see Gap 6, path f]; and no measurable genetic factor admits a clean subdivision between whites and African Americans; see also Taylor 2008b). On the other hand, racial group membership continues to bring disadvantages to African American individuals and, reciprocally, to bring benefits to white individuals (Flynn 2000, 142ff) (moderated somewhat, but in a diminishing set of circumstances, by affirmative action for African Americans). Putting both considerations together means that, if exposing the best way to ameliorate the effects of racial group membership for any individual leads us to seek empirical models of the heterogeneous pathways of development, all those pathways may have to factor in the common effect of membership in a racial group.

Suppose, give or take allowance for some common factors underlying the diverse pathways, we were to shift in focus from group membership to heterogeneous pathways. This move would come at the risk of bolstering a fiction that has gained currency in the United States, namely, that racial group membership no longer brings social benefits and costs. On the other hand, as long as researchers continue to track differences between averages for racial groups, they risk bolstering the ubiquitous stereotyping in which group membership is employed when deciding how to treat an individual. In sum, a genuine paradox that applies to the use of IQ test scores in US society seems to be that researchers and policy makers who want to move beyond explanations and policies based on racial group membership cannot escape taking into account the disadvantages and benefits individuals experience because of their group membership.

Additional puzzles

This last paradox invites wider-ranging historical and cultural inquiry spanning the three original puzzles: How do we account for the persistent interest in explaining differences among the averages for human groups, especially groups defined on racial grounds, in the United States, a country where public discourse emphasizes the freedom individuals should have to fully realize their unique potential? After all, the ranges within human racial groups for almost any trait are large and overlap with the ranges from other groups. Moreover, it is hard to make policy out of any finding about the differences between averages for groups *unless individuals are treated*—by teachers, social workers, medical practitioners, and so on—*on the basis of their group membership*.

We might generalize the issue that overlapping variation is discounted. When commentators do not address heterogeneity, are they making typological or essentialist assumptions? Has a racially essentialist imagination facilitated the transfer of conventional statistical tools from agricultural to human research? Might it be possible to pinpoint paths not taken or objections not taken up in scientific and public debates about group average differences? Could such blind spots be, in turn, interpreted in terms of the persistence of racial types as an organizing category in American social and scientific thought? Similarly, might the transfer of tools from selective breeding in agriculture and the laboratory to analysis of human variation speak to a persistence of unspoken eugenic hopes and fears? How are responsibility and causation conceived by people when they talk of individuals in terms of their group membership? Just who is empowered to do something as a result of an analysis of average group differences—and who is given license not to have to do anything? How have the answers to all these questions evolved over the last century?

Answers to the preceding questions could well affect our understanding of new molecular genetic techniques even though (see Item I) they may appear to circumvent the limitations of the classical methods of analysis of hereditary variation that have been the focus of Part I's Short Critique. The Introduction suggested that it is unwise to adopt the newer quantitative genetic techniques without understanding the conceptual and methodological gaps in the classical methods (Item B). A more critical understanding of classical methods allows researchers to evaluate whether similar or analogous oversights limit the progress that can be made with the newer methods (Items I and J). If we were to explore not only the gaps, but also the social and historical context in which the gaps have been obscured, would that cast doubt on any hope that molecular genetic analysis operates unconstrained by that context? In particular, can we expect the research trajectories followed today to be unmarked by previous pressures to explain differences among the averages for human groups defined on racial grounds (Shim et al. 2014)?

*   *   *   *   *

> It is one thing to criticize the methodology of specific studies. It is quite another to suggest… that we reject the results of an entire field of scientific inquiry. This might have been warranted for some pseudoscientific systems, such as astrology, alchemy, and the Ptolemaic astronomic system. It is highly unlikely that modern psychiatric genetics will be judged by future historians of science to be in such company (Kendler 2005, 10).

Suppose that critical commentators on heritability studies took up the suggestion made several paragraphs ago, namely, to revisit publications that have interpreted heritability and genetic variance as measuring the size of the influence of genetic factors on the process through which the trait develops. Suppose, after taking account of the eight gaps identified in Item B, they concluded that key results and interpretations from a century of quantitative genetics are not justified or, at best, are unreliable. What should they do?

Sociology and history of science remind us that critique is rarely sufficient for a dominant paradigm to be abandoned. Perhaps human quantitative genetics could be viewed, *contra* Kendler, as akin to alchemy, which was a field of inquiry that provided observations, questions, tools, debates, careers, and institutions that modern chemistry built on, but ultimately had to break away from, to make further progress. Our thinking about the status of quantitative genetics might then be informed by historians who have examined the shifts that led to the abandonment of alchemy by the early eighteenth century and to its eventual depiction as an exemplar of pseudoscience (Principe and Newman 2001, Principe 2007). Even if critical commentators do not take up the comparison to alchemy, they might consider what they should do if they think that scientists need to break away from fundamental and long-held assumptions and interpretations (Taylor 2011).

My personal response to this last puzzle is represented in the didactic impulse of Part I: *Here, readers, are foundations that, if you built on them, you would avoid the problems and confusions that beset heritability studies and their interpretation.* Part II continues in this vein, but extends the didactic work into a more technical discussion of what quantitative genetic models and causal claims would look like if researchers were to take all the eight gaps into account at all times.

PART II

CRITICAL, TECHNICAL: REWORKED MODELS AND CAUSAL CLAIMS

Introduction to Part II

The book's introduction noted that "[r]esearch into the genetics of complex traits has moved from the estimation of genetic variance in populations to the detection and identification of variants that are associated with or directly cause variation" (Visscher et al. 2007). This statement, however, takes as given what the Short Critique of Part I has called into question, namely, that estimation of so-called genetic variance is a meaningful and reliable basis for the newer research. The systematic presentation in Item B of the eight gaps was designed to show that, by and large, the methods of heritability studies cannot show anything clear and useful about genetic and environmental influences, especially in the case of human behaviors and other traits.

If we say *no* to nature-nurture research—that is, to heritability studies—how then should heredity and variation be studied? An obvious answer is explore methods that attempt to identify the specific, measurable genetic and environmental factors *without reference to the trait's heritability* or the other fractions of the total variance (an option mentioned in Item B under Gaps 3 and 6—Translation to hypotheses). Exploration of heritability-free methods is taken up in Part III. But there is some ground to be prepared first.

"If we say *no* to nature-nurture research..." begs an important question: "We" is not a given group; its membership has to be built up. If alternative directions of research into heredity and variation are to be productive, many contributors are needed. The Short Critique might have convinced *some* researchers and commentators on heritability studies to leave behind long-held assumptions and interpretations and to explore alternatives. (If that is the case for you, you might skip Part II and go directly to Part III.) However, reserve or resistance can be expected from other readers. After all, the methods used in estimating heritability and other fractions of variance have formed the basis of careers, fields, software packages, and policy positions, as well as of thousands of publications subject to extensive debate. How, it might be thought, could all those involved have gone wrong and for so long? Moreover, even if classical quantitative genetics has shortcomings, what is the alternative—what can researchers use in place of the highly elaborated infrastructure of data analysis (Falconer and Mackay 1996; Lynch and Walsh 1998; Holland et al. 2003)?

To address these skeptical questions, Part II turns to matters that are more technical, as is evident in the use of equations and some tables and figures. The phrase "all those involved have gone wrong" in the first of these questions might seem to prefigure that this technical discussion would take the debating form: "X and Y say this, but under *my* account that should be corrected to say..." This is not, however, Part II's primary expository mode. Before such an exchange on a given technical or empirical issue would be of any benefit, it would be important to understand how the first gap has been addressed. Are distinct multiple meanings of key terms, such as *genetic*, clear? If so, how is the second gap addressed? And so on—these understandings would need to be clarified for each of the eight gaps.

For example, consider the heterogeneity-centered problems introduced in Puzzle 2 and discussed under Gap 6. I have been asked how these problems relate to other problems confronted in heritability studies, say, so-called genotype-environment interaction. The question cannot be answered, however, unless I know whether the questioner's conceptualization of genotype-environment interaction keeps summaries of variation among traits distinct from measurable factors that underlie these traits (Gap 2). Indeed, in my reading, as noted in the Introduction, "[p]revious researchers and commentators have either not acknowledged the gaps, not developed the appropriate responses, or not consistently sustained those responses."

Instead of an author-specific critique, Part II explores what quantitative genetics would look like if Gap 2—between fractions of variation in a trait and underlying measurable genetic factors—were preserved throughout, from the derivation of formulas for analysis of data on similarities among relatives to the interpretation of partitioning of variation in terms of causal factors. The exploration begins by taking genealogical relatedness into account without using the standard models of quantitative genetics that refer to theoretical genes (Item D). Alternatives to some standard assumptions of classical quantitative genetics can be described using *gene-free formulations* and shown to have nontrivial implications for interpreting the partitioning of trait variation into components (Item E). Even if shortcomings in the standard approach to classical quantitative genetics were to be overcome, which would require further data collection and analysis, it is only in a highly circumscribed sense that partitioning of variation sheds light on causal factors. I call this circumscribed sense of causality *rerun predictability* (Item F).

Items D–F are critical then in the sense of understanding ideas better by holding them in tension with alternatives (Taylor 2008c). These items are also critical in the sense of amplifying the gaps of Part I and the conclusion that the methods of heritability studies cannot provide clear and useful information about genetic and environmental influences. Item G notes, however, that in agricultural research, hypotheses about underlying genetic and environmental factors can be derived using partitioning of variation *provided varieties can be grouped to reduce the likelihood of underlying heterogeneity*. If we consider ways that people address or suppress the possibility of underlying heterogeneity, a more general principle emerges: We may try to address our limited knowledge about the causal dynamics of a certain phenomenon through further research, but whether it is necessary to do so depends also on the actions possible or proposed on the basis of what is known (or unknown). Combining this concept of causality grounded in practice with rerun predictability allows Item G to resolve the deep puzzle, raised under Puzzle 3 in Item A, about the steps and assumptions through which heritability and other quantities that summarize the variation among measurements of a trait made *at one point in time* are supposed to shed light on the influence of underlying measurable factors involved in the *processes* of reproductive transmission and development of the trait.

Of course, drawing attention to the gap between analysis of traits and identifying relevant underlying factors is by no means a new point to make, but the distinction has not always been upheld, even by critical commentators. Reworkings of accounts by Lewontin, Turkheimer, and others are presented in Item H, leading me ultimately to suggest that researchers really want to be talking about the genetic factors and perhaps environmental factors influencing the development of the trait in question. Part III, therefore, will leave heritability studies behind and explore instead research on specific, measurable genetic and environmental factors underlying human traits. Certain themes developed in Part II, concerning heterogeneity and causality grounded in practice or action, will carry over and shape that exploration.

Item D. Gene-free formulation of classical quantitative genetics

In which genealogical relatedness is taken into account without using the standard models of quantitative genetics that refer to theoretical genes. The resulting formulation for the analysis of variation of a trait makes it harder to conflate that variation with variation in the genetic and environmental factors underlying the trait.

1. A notation that respects key distinctions (Gaps 1 & 2)

Ambiguous terms (see Gap 1) and shorthand names (see Gap 2, a.) help lead to the common but unjustified interpretation of the partitioning of a trait's variation into different components in terms of "the relative contributions of differences in genetic and non-genetic factors" (e.g., Wikipedia n.d., a). This conceptual slippage from traits to factors influencing the development of the trait needs a strong antidote, which is provided by the *gene-free formulation* developed in this item.

First, recall that the terms *variety* and *location* serve in this book as replacements for the ambiguous terms *genotype* and *environment* (see Gap 1). A variety is a group of individuals whose relatedness by genealogy can be characterized, such as offspring of a given pair of parents, or a group of individuals whose mix of genetic factors can be replicated, as in an open pollinated plant variety or pure (genetically identical) line. A location is the situation or place in which the variety is raised, such as a family or a specific experimental research station.

Second, the term variety-location interaction will be used here as a synonym only in the classical quantitative genetic sense of genotype-environment interaction (see equations 1 and 8 below). In everyday terms, a high degree of variety-location interaction simply means that the responses of the observed varieties across the range of the observed locations do not parallel one another. That is, one variety may be highest for the trait in one location, but another variety may be highest in another location—or, at least, the difference between any two varieties may change substantially from location to location. This sense of genotype-environment interaction is distinct from the use of the same term (or, synonymously, *gene-environment interaction*) for situations in which *genotype* denotes a value of a measured genetic factor, *environment* denotes a value of a measured environmental factor, and an *interaction* means that the quantitative relation between the trait and one of the factors varies according to the measured value of the other factor (e.g., Moffitt et al. 2005).

Now, consider the general case of an agricultural evaluation trial where it is possible to observe a set of animal or plant varieties in each of a set of locations, and to raise replicates for each variety-location combination. If the trait is recorded in all replicates, the data can be fitted to an *additive* or *linear* model that connects the values of the trait for an individual to the summation of several contributions:

$$y_{ijk} = \qquad m \qquad +v_i \qquad +l_j \qquad +vl_{ij} \qquad +e_{ijk} \qquad\qquad (1)$$

where y_{ijk} denotes the measured trait y for the ith variety in the jth location
    and k^{th} replication;

m is a base level for the trait;

v_i is the additional contribution of the ith variety;

l_j is the additional contribution of the jth location;

vl_{ij} is the additional contribution from the i,j^{th} variety-location
    combination—in statistical terms, the variety-location-interaction
    contribution; and

e_{ijk} is an unsystematic or noise contribution adding to the trait measurement.

(The term *contribution* is used in place of the technical term *effect* because the latter has an everyday connotation of the influence of some causal factor. Such a connotation tends to be especially confusing in the case of so-called shared versus nonshared environmental effects; Turkheimer 2000 and Item H.3.)

Any additive model like equation 1 can be converted to a model that adds up variances related to these contributions. (Recall: variance is the common measure of variation.) Such a model allows the variance of the trait to be partitioned into these component variances (i.e., an analysis of variance or ANOVA). The conversion of equation 1 to variances is made as follows: If each kind of contribution is uncorrelated with any of the others, and m is set at the average of the trait over all varieties, locations, and replicates, then the average of each of the other contributions is zero. If m is subtracted from both sides of equation 1, which are then squared and divided by the total number of individuals to arrive at the average of these squared contributions, the result is the following partitioning of variance:

$$Y \quad = \quad V \quad + L \quad +VL \quad +E \tag{2}$$

where

Y denotes the variance of the y_{ijk} observations as a whole,

V denotes the variance of the v_i terms, etc.

(This notation is used from here on: lower case letters with subscripts for contributions from specific varieties, locations, and replicates; upper case letters for variances corresponding to each contribution.) When both sides of equation 2 are divided by Y, we get the fractions of the overall variance summarized in Table D.1.

Table D.1. Fractions of variance of the observations as a whole

Symbols*	Source of variation	Alternative names for fraction
V/Y	Between-variety averages	Heritability (broad-sense) or h^2
L/Y	Between-location averages	Shared environmental effect
VL/Y	Between-variety-location-interaction averages	Genotype-environment interaction
E/Y	Noise	Error; Residual; Between replicates

* See text for definitions of symbols

Consider how equation 1 can be fitted to the data from the agricultural evaluation trial (as illustrated schematically in Figure B.1). To use the simplest case where the same number of replicates are observed for each variety-location combination, m is estimated by the average over all varieties, locations, and replicates; the estimate of v_i is the average of y_{ijk}'s for variety i across all the observed locations and replicates *minus* the estimate for m; similarly for the estimate of l_j. The estimate of vl_{ij} is the average of y_{ijk}'s for variety-location combination i-j after allowing for v_i , l_j , and m, that is, the average *minus* the estimates for the other three quantities.

(In practice, it is possible to partition the variance of the traits—to fit observations to, in this case, equation 2—without making explicit the original additive model—in this case, equation 1—and without estimating the values of the various contributions that fit the data. Nevertheless, the partitioning of variation can always be related back to an additive model of contributions.)

Equation 1 was introduced in the context of a trial where a set of animal or plant varieties is grown in every one of a set of locations. However, all the variances in equation 2 can still be estimated even if observations are made only in a subset of all the variety-location combinations *as long as the subset is randomly chosen from the range of possibilities.* (Of course, smaller subsets produce more variable estimates. That is to say, if the estimation were repeated over a number of subsets, the results would vary from subset to subset.)

The observations of the trait for a set of varieties and locations might also be divided into *classes* in various ways. For example, a class might be defined by all observations in which the variety is the same. In that case, equation 1 shows that, within the class of ith variety, m and v_i are constant, but the other terms vary. The expected average for the trait in question for the class is m + v_i (because the average in the class of each of the other contributions is zero), and the variance of those averages across classes is V. Similarly, for classes defined by all observations in which the location is the same, the variance of the averages for the classes is L. For classes defined by all observations in which both the variety and the location are the same, the variance of the averages for the classes is V + L + VL.

When the variances of the contributions that do not vary within the class (and are thus included in the class averages) are compared to the overall variance, Y, the results are what are called *intraclass correlations.* (Although intraclass correlations are a ratio of variances, the label correlation makes sense if it is noted that for classes of size two, this quantity is mathematically equivalent to the usual linear correlation of the two values when the order in each pair is arbitrary, as would be the case if one wanted to know the correlation, say, of heights in same-sex couples; Howell 2002.) Finally, if the sum of the first two intraclass correlations is subtracted from the third, the result would be VL, and if the last intraclass correlation is subtracted from 1, the result would be E. With intraclass correlation denoted by I and the subscript specifying the class, the following equations summarize this paragraph:

$$V / Y \qquad\qquad = I_V \qquad\qquad\qquad\qquad\qquad\qquad\qquad\qquad (3)$$

$$L / Y \qquad\qquad = I_L \qquad\qquad\qquad\qquad\qquad\qquad\qquad\qquad (4)$$

$$(V + VL + L) / Y \quad = I_{(V,L,VL)} \qquad\qquad\qquad\qquad\qquad\qquad (5)$$

Equations 2-5 imply that

$$VL / Y \qquad = I_{(V,L,VL)} \qquad\quad - I_V \quad\; - I_L \qquad\qquad\qquad (6)$$

$$E/Y \qquad\quad = 1 \qquad\quad - I_{(V,L,VL)} \qquad\qquad\qquad\qquad\qquad (7)$$

It is possible to estimate these intraclass correlations even if observations are made only in a subset of all the variety-location combinations, just as it is possible to partition the variance in the trait according to equation 2 in such subsets. Again, the subsets need to be randomly chosen from the full data set, ensuring no *systematic* differences between classes in the contributions from varieties, locations, variety-location combinations, and noise. Finally, there are statistical techniques that estimate the variances without calculating intraclass correlations, but the form of equations like 3 to 7 is used here to allow readers without training in statistical theory to appreciate the conceptual points.

2. The basic case of gene-free quantitative genetics

The analysis or partitioning of trait variation in the preceding formulation neither *requires* knowledge of nor, on its own, *produces* knowledge about the genetic or environmental factors that influence the trait (or "phenotype") in the various variety-location (or "genotype-environment") combinations. (This is also the case when path analysis and Structural Equation Modeling are used to partition the variation in a trait; see Item F.5). However, in estimating the fractions of the overall variance using equation 2 or using intraclass correlations, we have yet to take into account the genealogical relatedness between varieties.

Consider, therefore, a special case of the agricultural evaluation trial in which the varieties are replicated as twins or as relatives of some other known relatedness (e.g., half-siblings, cousins). Conceptually, the simplest analysis of variation in traits between relatives involves a comparison of three classes: monozygotic (MZ) twins raised apart (i.e., in randomly chosen locations); unrelated varieties (for humans: individuals) raised together within the same location (for humans: family); and MZ twins raised together (i.e., both members of each pair in the same location). Denoting these classes as MZA, UVT, and MZT, respectively, the formulas for intraclass correlations (derived again by identifying the contributions that do not vary within the replicates, i.e., twin pairs) are re-expressions of equations 3-7:

$$V /Y \qquad = I_{MZA} \qquad\qquad\qquad (3')$$

$$L / Y \qquad = I_{UVT} \qquad\qquad\qquad (4')$$

$$(V + VL + L) /Y \quad = I_{MZT} \qquad\qquad\qquad (5')$$

Equations 2 and 3'-5' imply that

$$VL /Y \qquad = I_{MZT} - I_{MZA} - I_{UVT} \qquad\qquad (6')$$

$$E/Y \qquad = 1 \quad - I_{MZT} \qquad\qquad\qquad (7')$$

To bring other classes, such as dizygotic (DZ) twins, into the picture, an elaboration of equation 1 is needed:

$$y_{ijk} = \qquad m \qquad + v^-_I \quad + t_{ik} \quad + l_j \quad + vl^-_{ij} \quad + tl_{ijk} \quad + e_{ijk} \qquad (8)$$

where y_{ijk}, m, l_j, e_{ijk} are as before;

t_{ik} denotes an additional contribution from the kth twin (replicate) in the ith variety;

tl_{ijk} is an additional contribution from the kth twin in the i,j^{th} variety-location combination;

v^-_i and vl^-_{ij} replace the v_i and vl_{ij} contributions in equation 1. (The superscript indicates that the new contributions would tend to be smaller given that t_{ik} and tl_{ijk} contribute to differences between twins [replicates].)

The partitioning of variance corresponding to equation 8 is:

$$Y = \qquad V^- \qquad +T \quad + L \quad + VL^- \quad + TL \quad +E \qquad (9)$$

where V^- denotes the overall variance of the v_i^- terms, etc.

Noting that for MZ twins $t_{i1} = t_{i2}$ and $tl_{ij1} = tl_{ij2}$, the intraclass correlations for the MZT and DZT classes are:

$$(V^- +T \quad + L \quad + VL^- \quad + TL \quad) / Y= I_{MZT} \qquad (10)$$

$$(V^- \qquad + L \quad + VL^- \qquad\qquad) / Y= I_{DZT} \qquad (11)$$

Note that $V^- +T = V$ and $VL^- + TL = VL$. Define a parameter, γ:

$$\gamma \quad = \quad (V^- + VL^-) \, / \, (V + VL) \tag{12}$$

or, equivalently,

$$1 - \gamma \quad = \quad (T + TL) \, / \, (V + VL) \tag{13}$$

The value of γ can be empirically determined if data from all three classes MZT, DZT, and UVT are available. Based on equations 4', 10 and 11:

$$\gamma \quad = \quad (I_{DZT} - I_{UVT})/(I_{MZT} - I_{UVT}) \tag{14}$$

If V^+ is used in place of $(V + VL)$ (for reasons that emerge in Item E), equations 10 and 11 can be rearranged to yield:

$$V^+ / Y \quad = (I_{MZT} - I_{DZT}) \, / \, (1 - \gamma) \tag{15}$$

$$L / Y \quad = (I_{DZT} - \gamma \, I_{MZT}) \, / \, (1 - \gamma) \tag{16}$$

Equivalent parameters and equations can be formulated for classes of relatives other than twins (Box D). However, the preceding formulation suffices as a proof of principle: gene-free analyses of variation can be formulated that take degree of relatedness into account through a parameter to be empirically determined. (Table D.1 summarizes examples of gene-free analyses for twins.)

Box D. Gene-free formulations for classes of relatives other than twins

Gene-free formulations can be derived and applied through five steps:

1. Define varieties in terms of the progenitors for individuals in the variety, e.g., a clone, a pair of parents, one mother and unrelated fathers, a pair of grandparents.

2. Specify the variant of equation 8 that encompasses the different kinds of relatives possible for such progenitors (e.g., siblings, first cousins).

3. Spell out the intraclass correlation equations for the different kinds of relatives in the different circumstances (e.g., raised together, raised apart).

4. Rearrange the intraclass correlation equations to produce estimators for the variance fractions and for any empirically determined parameter that had to be introduced to take degree of relatedness into account.

5. Collect data for the classes of relatives needed in order to estimate the variance fractions and parameters of interest using the equations from step 4.

Table D.1. Gene-free analyses for twins

Data Set & Assumption	Values that can be estimated*				Equations
1. MZT—each twin in a pair raised in the same location	$V^+ + L$			E	5', 7'
2. MZT, DZT (assuming some value of γ, not necessarily = .5)	V^+		L	E	15, 4' or 16, 7'
3. MZT, DZT, UVT in the same population (with actual value of γ estimated)	V^+		L	E	14, 15, 4' or 16, 7'
4. MZT, MZA, UVT in the same population	V	VL	L	E	3', 4', 6', 7'
5. MZT, DZA, DZT, UVT in the same population (and assuming that V^- / VL^- and T / TL are equal)	V	VL	L	E	12, 13, 14, 15, 4' or 16, 7'; also $V^- / Y = I_{DZA}$

* Estimated as fractions of Y. V^+ denotes V + VL

3. A numerical illustration of the gene-free formulation

The terms in which the gene-free formulation has been introduced are not dependent on any specific data set. Nevertheless, for readers who want to see how this formulation plays out in numbers, this section provides a numerical illustration of the gene-free formulas applied to one of the simulated data sets from Appendix 1. (Appendix 1 covers a wide range of values of broad-sense heritability, V/Y, and other fractions of variance. It also contrasts estimates of fractions of variance given by equations 15 and 16 with those from equations 18 and 19, which are to be introduced in Item E.)

First, equation 8 is used to generate simulated observations for monozygotic and dizygotic twin pairs for a set of 100 varieties raised together in each of a set of 100 locations (MZT and DZT), for twin pairs raised in different locations (MZA), and for unrelated varieties raised together (UVT). Each observation derives from the same pre-set values of the fractions of variance. The relevant equations from the gene-free formulation are then applied to the complete data set to estimate the values used in generating the observations (Table D.2). Given the random process to generate these observations, the estimates should not be identical to the pre-set values. There is, however, a close match, with the greatest discrepancy evident in the estimate of γ.

Table D.2. A comparison of the original values and estimates applying intraclass correlation formulas to the full data set

Quantity	Value used in generating the observations	Intraclass correlation formula	Equation number	Estimate
V/Y	.4	I_{MZA}	3	.41
L/Y	.2	I_{UVT}	4	.21
VL/Y	.2	$I_{MZT}-I_{MZA}-I_{UVT}$	6	.19
E/Y	.2	$1-I_{MZT}$	7	.19
γ	.3	$(I_{DZT}-I_{UVT}) / (I_{MZT}-I_{UVT})$	14	.25

Next, equations 15 and 16 are applied to the complete data set to produce estimates of V^{+} and L as fractions of Y (Table D.3). The estimates also match the values corresponding to those used in generating the observations, but not as closely as in Table D.2 (which is not surprising, given that γ, the least-close match above, factors into the equations).

Table D.3. A comparison of the original values and estimates applying equations 15 and 16 to the full data set

Quantity estimated	Value corresponding to those generating the observations	Intraclass correlation formula	Equation number	Estimate
$V^{+}/Y =$ (V+VL)/Y	.6 = .4 + .2	$(I_{MZT} - I_{DZT}) / (1-\gamma)$	15	.57
L/Y	.2	$(I_{DZT} - \gamma I_{MZT}) / (1-\gamma)$	16	.23

Finally, to mimic the situation found in human twin studies, subsets of the full data set are randomly sampled so that a given MZ variety is observed only for one location in the MZT sample, similarly for the DZT sample, and varieties and locations for the MZT and DZT samples do not overlap. Table D.4 presents results for samples of 48 MZT pairs, 48 DZT pairs and 48 UVT pairs, with the sampling process repeated 100 times. The actual value lies within a small interval around the estimate (less than 1 standard deviation), but the match for the samples is not as close as in Table D.3. This would be expected given that only a subset of the complete data set is used.

Table D.4. A comparison of the original values and estimates applying equations 15 and 16 to samples from the full data set

Quantity estimated	Value corresponding to those generating the observations	Intraclass correlation formulas	Equation number	Estimate	Standard deviation
$V^+/Y =$ $(V+VL)/Y$	.6	$(I_{MZT} - I_{DZT})$ $/ (1-\gamma)$	15	.52	(.14)
L/Y	.2	$(I_{DZT} - \gamma I_{MZT})$ $/ (1-\gamma)$	16	.24	(.14)

4. Technical conditions and data availability

The partitioning of variance in the preceding sections depends on certain technical conditions:

a) An **agreed-on scale** has been used for measuring the trait and calculating the variance to be partitioned into fractions.

b) If not all variety-location combinations are represented in the data—as is obviously the case for human data sets—the **sampling of combinations is random** or, at least, produces no variety-location ("genotype-environment") correlations. Moreover, there are no systematic differences between the varieties and locations in which MZ pairs are observed and those in which DZ pairs are observed.

c) **Zygosity** of twins (i.e., monozygotic versus dizygotic) is correctly ascertained and representatively sampled. (Similarly for classes of relatives other than twins.)

d) The **same overall variance** of the trait applies in any of the classes as in the general case, e.g., Y in equations 10 and 11 refers to the same quantity.

e) Estimates are given with standard deviations, confidence intervals, or other assessments of their **statistical uncertainty**.

f) **Statistical bias** in the estimation of the parameters and fractions of variance is not substantial or is corrected for (Donoghue and Collins 1990).

g) **Equal environment**: For the different classes (e.g., MZT, DZT, and UVT), the treatment or experience of members of a class within a location is unaffected by their degree of relatedness (e.g., whether they are MZ twins, DZ twins, or unrelated).

Whether these conditions can be met in human situations, or whether it matters when they are not, are issues of ongoing controversy (e.g., Richardson and Norgate 2005; for accessible reviews, see Nuffield Council on Bioethics 2002, Parens and Chapman 2006). For example, Bouchard and McGue (1981) use estimates of intraclass correlations for human IQ test scores averaged from many studies; standard deviations or confidence intervals are not shown. The averaging clearly does not meet condition d., let alone show readers that the technical conditions applied in each of the cases that went into the average.

In agricultural situations the technical conditions can, in principle, be met; data sets that allow estimation of components of variance using equations 3'–7' are routine. At the same time, data have not yet been collected with a view to using the gene-free formulation of quantitative genetics to compare classes of defined degrees of relatedness. Indeed, a considerable effort would be required to use the steps of Box D to spell out equations that encompass the variety of degrees of relatedness addressed in classical quantitative genetics (e.g., Holland 2003). In this era of molecular genetics it may be hard to find the resources and interest for such an undertaking. Nevertheless, the gene-free formulation as introduced in this item allows for a fresh look at classical quantitative genetics. As Item E will show, the implications for genetic partitioning of trait variation into components and interpreting the results are not trivial.

Item E. The results and interpretation of classical quantitative genetics under alternatives to three standard assumptions

In which three assumptions involved in standard quantitative genetics practice are described, plausible alternatives are identified, and the implications of the contrasting assumptions for interpreting classical quantitative genetic partitioning of trait variation are drawn.

(Note: These assumptions have been previously identified in quantitative genetics or critical commentaries, but the account here makes use of the gene-free formulation of classical quantitative genetics from Item D.)

Assumption 1. Partitioning of trait variation into components requires models of theoretical genes with simple Mendelian inheritance and direct contributions to the trait.

As an example of this assumption, consider the common ADCE model for human twin studies (e.g., Feng et al. 2009), which can be expressed as:

$$y_{ijk} = \quad m + \quad a_{ik} \quad + d_{ik} \quad + l_j \quad + e_{ijk} \tag{17}$$

where y_{ijk}, m, l_j, e_{ijk} are as before (with l replacing c for consistency with the terminology of Item D);

a_{ik} denotes an "additive" genetic contribution from the kth twin in the ith variety;

d_{ik} denotes a "dominance" genetic contribution from the kth twin in the ith variety;

with $i = j$ if the twins are raised in the same location.

(In the corresponding partitioning of variance, A/Y estimates heritability in the *narrow sense*.)

The terms *additive* and *dominance* derive from the construction of the standard models of quantitative genetics (Falconer and Mackay 1996; Lynch and Walsh 1998), which builds from the case of a trait governed by a pair of alleles of a single gene (i.e., at a single locus) where all the individuals are raised in a single location. In that location, the genes contribute directly to the trait in the sense that the presence or level of such a trait depends only on whether the individual has two copies of one allele (i.e., is *homozygote* for that allele), two of the other, or one of each (*heterozygote*). Dominance refers to the degree that heterozygous individuals depart from the intermediate between the two homozygous forms. The variance in the single-locus trait in a given location depends on the difference between the homozygous forms, the degree of dominance, and the frequency of the different alleles.

For example, phenylketonuria (PKU) in humans is associated with having two copies of a malfunctioning allele for the enzyme phenylalanine hydroxylase (PAH). The cognitive development of such individuals is extremely impaired by the level of phenylalanine present in normal diets. The heterozygote condition generally goes undetected (but see Alós et al. 1993); in other words, the functioning allele is completely dominant. In a location defined by access only to normal diets, relatives would resemble each other across multiple traits more than unrelated individuals would. The greater resemblance follows because if, say, a twin has PKU, both parents have at least one copy of the malfunctioning PAH allele, so the other twin is more likely to have two malfunctioning PAH alleles than is an unrelated individual (i.e., one chosen at random from the population). Similarity in cognitive impairment follows directly. In a location defined by access to a special low-phenylalanine diet and adherence to that diet by individuals who are homozygous for the malfunctioning allele, relatives would resemble each other only a little more than unrelated individuals for most traits, such as IQ test scores. They would still resemble each other more than unrelated individuals in relation to the ability to metabolize phenylalanine (Paul and Brosco 2013).

For a trait governed by alleles at a single locus, the resemblance of relatives for the given trait is quantified by the intraclass correlation. For MZ twins this is 1—there is no variation within the pairs—but for DZ twins or pairs of other relatedness, the resemblance varies. (Recent research shows discordance between MZ twins at the genetic level, e.g., Bruder et al. 2008, but the simplification of an intraclass correlation of 1 for MZ twins is preserved in this book.) For example, in the case of PKU in locations where the special diet is not available, a proportion of functioning PAH in human populations greater than .98

translates to an intraclass correlation of DZ twins and full siblings close to 1/4.

Because few traits are dictated only by alleles at a single locus, the standard models of quantitative genetics envision the influence of alleles at many loci adding up to shape the traits, and thus to contribute to the variances of trait values that are to be analyzed. The models allow for some noise (from measurement error or unsystematic variation between the replicates of the variety) and for the variety to be raised in a number of locations (by incorporating into the equations a term for variance across locations of the average value of the trait in each location). Variety-location interaction as well as variety-location correlation can also be included in the models (Lynch and Walsh 1998, 107ff).

An alternative to this first assumption is to analyze the variation in a trait without making reference to theoretical genes, their dominance relations at individual loci, or their summation over many loci. The gene-free analysis of variation (Item D) simply starts with the traits and their variance and incorporates an empirical parameter to take degree of relatedness into account.

Assumption 2. All other things being equal, similarity in traits for relatives is proportional to the number of genes common to the relatives as a fraction of the genes that vary in the population.

Under this second assumption, all other things being equal (i.e., equal location and noise contributions), DZ twins, for example, would be half as similar as MZ twins because MZ twins share all their genes, while DZ twins share, on average, half of the genes that vary in the population (as stated, e.g., by Kendler and Prescott 2006, 42). (The common statement that DZ twins share half of their genes is potentially misleading given that it obscures the large proportion of genes shared by all members of any given species.) This assumption is equivalent to setting γ to .5 in equation 12, which means that equations 15 and 16 become:

$$V^+/Y \quad = 2\,(I_{MZT} - I_{DZT}) \tag{18}$$

$$L/Y \quad = 2\,I_{DZT} - I_{MZT} \tag{19}$$

Equations 18 and 19 resemble the standard formulas using data from MZ and DZ twins to estimate broad-sense heritability and the fraction of variance for locations, e.g., Rijsdijk and Sham (2002), *except* that equation 18 refers to $V^+/$ Y, whereas heritability is, by definition, V/ Y (a discrepancy to be discussed under Assumption 3). Assumption 2 is also reflected in the construction of most path diagrams and Structural Equation Models in quantitative genetics (Lynch and Walsh 1998, 823ff).

One alternative to this second standard assumption is using an empirical determination of parameters to account for defined degrees of relatedness. For example, γ can be estimated using equation 14 and the fractions of variance for V^+ and L can be estimated using equations 15 and 16 instead of equations 18 and 19. (Similarly, for parameters in equivalent equations when classes of relatives other than twins are studied; see Box D.)

Assumption 3. In analyses of human data, variety-location-interaction variance ("genotype-environment interaction" variance) can be discounted.

First, recall from Item D.1 that interaction variance refers here to the classical quantitative genetics sense of the term. (The conceptually distinct interactions between measured genetic and environmental factors are *not* discounted in human studies; for an entry point, see Moffitt et al. 2005 and Item I.2.) The assumption that variety-location-interaction variance can be discounted enters when the interaction terms are omitted from the additive model for partitioning of variance (e.g., the ACDE model) or when V^+/Y in equations 15 or 18 is taken to estimate V/Y.

The origins of the third assumption can be seen in the single-locus case used as the theoretical starting point for the standard models of quantitative genetics (see Assumption 1 above). Heritability in the single-locus case is *within-location* heritability, and, as such, is related to $V^+/$ Y, not V/ Y. To visualize why this is so, consider equation 1 restricted to some given location j. In that location, m $+l_j$ is constant across all observations, but $v_i + vl_{ij}$ varies with i (the variety), so the variance of these within-location variety contributions is V + VL. This feature of the standard model is not altered by the subsequent incorporation into the model of terms for noise and location variance.

An alternative to discounting variety-location interaction is to include interaction terms in the additive model and then to secure the kinds of data needed to estimate the separate fractions of variance associated with V and VL. (For example, for twins, equations 15 and 18 cannot separate the V and VL fractions, but equations 3' and 6' can, provided data are available for MZT, MZA, and UVT. Given that data for MZA in humans are rare, data on DZA and UVT from adoption studies could be used instead provided it is assumed that V^- / VL^- and T / TL are equal; see Table D.1.)

4. Implications of gene-free formulations and contrasting assumptions for classical quantitative genetics partitioning of trait variation

The gene-free formulation and alternatives to the standard assumptions have key implications for interpreting any classical quantitative genetics partitioning of trait variation into components, and thus for subsequent research and applications based on those interpretations. Again, some of the issues in this section have been raised before (as have other important issues, such as the technical conditions listed in Item D.4), so the discussion here is focused on what follows from the gene-free formulation and the alternative assumptions. (To be clear: the purpose is to elaborate on the gene-free formulation, not to argue that these issues can be raised *only* using that formulation.) The implications are drawn at a theoretical level and are not dependent on empirical data from any particular situation. (Possible implications at an empirical level can be seen in the data sets in Appendix 1.)

4.1 Gene-free quantitative genetics

a) Distinction between traits and underlying measurable factors. As noted previously, the quantitative genetics term *heritability* is commonly described in terms like the "contribution of genetic differences to observed differences among individuals" (Plomin et al. 1997, 83) or the "fraction of the variance of a phenotypic trait in a given population caused by (or attributable to) genetic differences" (Layzer 1974, 1259). In such descriptions, the distinction between traits and underlying measurable factors is not kept clear (Gap 2 in Item B). The gene-free formulation, by using models that make no reference to theoretical genes, serves as a reminder that translation from descriptive quantitative genetics analyses of traits to hypotheses about causal factors is far from direct (Gaps 3, 6, and 7). The difficulty of translation to hypotheses about measurable factors also applies to the fraction of trait variation for between-location means and for interaction, i.e., between means for variety-location combinations. (See also Item H.3 for an interpretation of residual variance as "non-shared environmental effects.")

To remark on the distinction is not new, but it is often obscured by the language used, as evidenced in the quotes above, or by combining in one analysis measurable factors and variation between varieties and locations (Item H). (Keeping the distinction between traits and underlying measurable factors well marked is a good reason for using the agricultural terms *variety* and *location* in place of *genotype* and *environment*; see Gap 1 in Item B.)

b) Interpretations based on contributions of theoretical genes not necessary. Given the existence of a formulation of classical quantitative genetics free of reference to theoretical genes (Item D), data need not be interpreted with reference to the contribution of differences in unknown genetic factors or their interactions. In particular, literal readings are not required for additive models that refer to dominance and nonadditive genetic variance (e.g., equation 17). Moreover, although narrow-sense heritability will be less than broad-sense heritability, we do not have to ascribe that difference to a fraction of the variance reflecting dominance relations at individual loci of theoretical genes. Now, it may turn out in practice that formulas for narrow-sense heritability provide more accurate predictions of change in a trait under artificial selection than do formulas for broad-sense heritability, but establishing that outcome is an empirical matter. It is not something resolvable by reference to quantitative genetic models based on theoretical genes.

4.2 Estimation and interpretation of similarity in traits for relatives

a) Similarity between relatives determined empirically. Other than in trivial situations, the use of the gene-free formulation to compare classes of defined degrees of relatedness requires parameters such as γ in the case of MZ versus DZ twins. These parameters can be determined empirically, provided the appropriate classes of data are available (Box D and Table D.1) and provided the necessary technical conditions have been shown to hold (Item D.4). The empirically determined values may depart from the commonly used heuristic values. Under the standard assumptions, observed departures from heuristic values have led to interpretations in terms of dominance relations at individual loci or nonadditivity of the summation over many loci. However, as noted above (section 4.1b), the availability of a formulation of classical quantitative genetics free of reference to (theoretical) genes implies that data need not be interpreted with reference to the contribution of differences in actual (unknown) genetic factors or their interactions.

b) Relevant correlations based on observed traits, not the proportion of shared genes that influence the development of those traits. One virtue of the empirical determination is being able to dispense with the second standard assumption, which is not a reliable heuristic. This unreliability can be shown by using plausible models of the contributions of multiple genes to a trait and finding, all other things being equal, ratios of DZ similarity to MZ similarity that are not .5 and that vary considerably around their average. Consider, for example, a disease trait modeled in the following way: The trait occurs when the combined "dosage" from many loci exceeds a threshold, where each pair of alleles contributes a full, zero, or half dose according to whether the alleles are, respectively, both the same for one variant, both the same for the other, or one of each. In this case, the intraclass correlation varies according to the frequency of alleles, level of dominance, and so on. The varying values are mostly above .5 even when there is no dominance (as summarized in Appendix 2).

Of course, it is possible to put forward more complicated models of the interaction of genes and the timing of their influence during development. The point here does not, however, depend on the validity of the model in the previous paragraph or of any particular hypothetical model of multiple genes contributing to the trait. The reason that the standard assumption is unreliable is simply that the relevant correlations need to be based on observed *traits* and, as such, cannot be directly drawn from the proportion of shared *genes* that influence the development of those traits. (This assertion is not affected by researchers now being able to determine empirically the exact proportion of genes shared for particular relatives in a given population; Visscher et al. 2006.) For the same reason, heuristic values of the similarity of relatives of other degrees (half-siblings, cousins, etc.), which are ubiquitous in classical quantitative genetics, are also unreliable.

c) Possible research program for estimation of empirical parameters in agricultural situations and extrapolation to human situations. What average value and range would empirically determined values for genealogical relatedness have in agricultural and laboratory populations, where empirical estimation is not difficult? In studies of twins in such populations, is the value of γ generally close to .5? How widely do the values vary? If the average value and ranges for agricultural and laboratory populations are extrapolated to human quantitative genetics, what adjustment is needed to previously reported results from comparing MZ versus DZ twins raised together? The same questions could be asked of the equivalent values that arise in gene-free derivations of analyses for comparing relatives of other degrees of relatedness. Is there a linear relation between the empirically determined values for relatives of various degrees and the fraction of variable genes that each kind of relative shares? Perhaps a research program based on these questions would show consistent patterns; perhaps results would depend on the specific trait for the specific species for the specific locations in which the individuals were raised. That cannot be known until the investigations are carried out.

4.3 Variety-location-interaction variance estimated, not discounted

Variety-location interaction variance is routinely estimated in agricultural studies. (Recall, from Item D, that interaction variance refers here to the classical quantitative genetics sense of the term.) A significant interaction variance tempers any recommendations to farmers to adopt a certain variety that has a high average across locations for a desired trait. Having a set of varieties observed in each of a set of locations makes it straightforward to estimate V and VL separately (performing an ANOVA based on equation 2 in Item D). In studies of human twins raised together, in contrast, each variety is observed in one location with two replicates (twins) for each of those variety-location combinations. Without collecting data for twins raised apart (see Table D.1) there is no way to ascertain how much the variation between varieties would change if the varieties were observed in locations (families) other than the ones in which they were actually observed, that is, there is no way to separate VL out of V^+.

How does interaction variance in human studies line up with the range of the levels found in the agricultural studies? (The large agricultural field trials mentioned in the introduction typically showed substantial interaction variance. The studies reviewed by Tabery [2014, 147-152] provide a more mixed picture.) The potential importance of interaction variance has long been noted (e.g., Layzer 1974, Lewontin 1974a, Plomin et al. 1977, Jacquard 1983; see Appendix 3 for contrasting formulations in mathematical notation). Yet the additive models used in partitioning variance in human quantitative genetics generally omit the interaction component, as evident in path diagrams (see Item F.5). Plomin et al. (1977), which is often cited in the context of low interaction variance for humans, considers only a proxy for variety-location interaction. For some trait, e.g., educational attainment, the statistical interaction of the average for biological parents with the average for adoptive parents is calculated. How well such proxy measures reflect actual variety-location interaction is, however, hard to assess in the absence of studies for a range of human traits in which the classes of data are collected that allow VL/Y to be separated out from V^+/ Y. (Tabery 2014, 152ff reviews for various human traits the evidence for low values of such proxy measures.)

If we do *not* assume that variety-location interaction variance can be discounted in human studies, two common claims become open to scrutiny: a) The effect of family members growing up in the same location (family) is of small importance; b) The trend for heritability estimates to increase over people's lifetimes is evidence that "genetic" differences come to eclipse "environmental" differences (Plomin 1999, C26). The first claim requires showing not only that the location variance is a small component of the total variance, but also that the variety-location-interaction variance is small (and thus V^+/Y close to V/Y). The second claim also requires showing that the variety-location-interaction variance is negligible; otherwise it could equally well be that it is the interaction component that increases over time.

4.4. Recapitulation: Three problem points in the standard derivation of quantitative genetics

This section follows a slightly different path to convey the implications of the three contrasting pairs of assumptions, but ends up affirming the previous sections as it identifies three *problem points* in the standard derivation of quantitative genetics.

As mentioned under Assumption 1, the standard derivation of quantitative genetics begins from a model of a trait governed by a pair of alleles at a single locus where all the individuals are raised in a single location (Falconer and Mackay 1996; Lynch and Walsh 1998). In other words, there are no contributions corresponding to differences between locations or to differences between replicates (i.e., noise from measurement error or unsystematic variation between the replicates of the variety) and V and VL are subsumed in a single term (as explained in Assumption 3). Let us call this combined term Y_S for the variance for a single locus in a single location. The intraclass correlation for MZ twins is 1, but for DZ twins the value varies with the degree of dominance and the frequency of the two alleles. Let us call this quantity g. (The difference, 1- g, corresponds to the average variance within the DZ twin pairs.)

Next, the standard models of quantitative genetics envision the influence of alleles at many loci adding up to shape the traits to be observed and analyzed. If each pair of alleles is assumed to add a small direct contribution to the trait—direct in the sense that each contribution is independent of the others—then the ratio of the variation between averages for DZ twins to the total variance is unchanged and the intraclass correlations remain the same. (Note that, because these intraclass correlations incorporate no environmental or unsystematic influences, they have been labeled as genetic similarity or genetic correlation. However, to avoid any risk of implying that similarity in traits as analyzed in classical quantitative genetics has a direct relation to similarity in genetic factors, the potentially misleading adjective genetic has not been used in the discussion of similarity in this book.)

Next, the models allow for some *noise*, which is assumed to be equal for both kinds of twins. In other words, E in equation 9 is the same when the equation refers to MZ twins as when it refers to DZ twins. Finally, when varieties are raised in a number of locations, the standard models incorporate a term for variance *across locations* of the average value of the trait in each location. Again, the models assume that this term, L, is equal for the different classes of relatives. In sum, for MZ twins, the variance between twin pairs has increased in steps from Y_S *to* Y_S +E *to* Y_S + L +E and, for DZ twins, from g Y_S + L +E *to* g Y_S + E *to* g Y_S + L +E.

Now, for a trait observed at a single location and governed only by alleles at a single locus, the intraclass correlations for DZ twins (or other classes of defined degrees of relatedness) can be determined exactly. However, to extrapolate from this and conclude that the same ratio holds for other kinds of traits requires *evidence* for the assumptions built into the subsequent steps in the derivation (e.g., each pair of alleles adds a small contribution independently of the others). Such evidence is lacking, which is not surprising given the problems inherent in trying to discriminate between the contributions of many different loci (Lewontin 1974b). This lack of evidence for assumptions is the first problem point in the standard derivation of quantitative genetics.

Suppose, however, that we put this problem point aside and adopt the final expressions above for the variance between twin pairs. The intraclass correlations for MZ and DZ twins are given by:

$$I_{MZT} \quad = \quad (Y_S \quad +L) / Y \tag{20}$$

$$I_{DZT} \quad = \quad (g\, Y_S \quad +L) / Y \tag{21}$$

These equations can be rearranged to yield estimates for Y_S and L that exactly match those given by equations 15 and 16 if g is seen as a synonym for γ and Y_S for V^+. The second problem point is that g or γ is not necessarily 1/2. And the third is that Y_S is V^+, not V; *the standard derivation has not separated V from VL*. As explained at the end of section 3 above, the location-specific variety-location-interaction variance is unavoidably bound together with the between-variety variance in the single-locus case; this combination is preserved through the subsequent steps of the derivation and thus into equations 20 and 21.

The three problem points can be removed by three steps: decoupling the estimation of quantities from any theory about summation of contributions at a single genetic locus; an empirical estimation of g or γ in place of the use of heuristic values based on the proportion of shared genes; and the inclusion of a separate term for VL followed by the collection of the classes of data needed to estimate V and VL separately. As indicated in Appendix 1, the difference this makes to empirical estimates depends on the values of V, VL, L, E, and γ, but may be substantial. To highlight the importance of overcoming the problem points, Taylor (2007) shows that an adjustment—albeit a simple one—that allows for a nonzero VL results in most human heritability estimates falling to values below the fractions for variance between-location-averages (the "shared environment effects").

Item F. Conditionality, repeatability, rerun predictability, and circumscribed causality

In which heritability and other fractions of the variation in the trait are subsumed under the general category of rerun predictability, a move that clarifies the limited causal significance of such quantities.

The gene-free formulation seeks patterns in the data for a trait, namely, to divide the variance into fractions that correspond to the different contributions included in a given additive model. This formulation makes no claims about the causal implications of the unknown genetic and environmental factors that underlie the traits. It merely adjusts the empirical estimation of the fractions and motivates questioning the common causal interpretations that others make from those fractions. Items F and G clarify conditions under which heritability studies can support causal claims, albeit limited ones.

I. Contributions are not properties of the varieties and locations alone

In heritability studies, the results of fitting observations to an additive model are conditional on the specific set of varieties and set of locations observed. (This conditionality is the case in *any* partitioning of variance or ANOVA.) The results are neither an indication of causes nor properties of the varieties that apply more generally (Item B, Gap 2, d.). One way to keep the conditionality in mind is to consider how v_i is fitted to the data. Recall (from Item D) the simplest case, when the number of replicates observed for every variety-location combination is the same: the estimated value of v_i is the average of y_{ijk}'s for variety i *across all the observed locations* and replicates minus the average over all varieties, locations, and replicates. In other words, the contribution v_i is not a property of the variety i on its own. Similarly, for any location contribution, the values that fit the data involve an average across all varieties.

A second way to visualize conditionality is to consider a specific set of varieties and locations, then observe how the estimates change as varieties or locations are omitted from the data. Table F.1 shows conditionality for Data Set 1, depicted in Figure F.1, in which four varieties are raised in two locations.

Table F.1 Estimates of contributions and fractions of variance for a simple data set

Data Set 1			Estimates of contributions		Variance & heritability estimates	
Location	1	2	m	3.0	V	0.3125 (13%**)
Variety			l_1	1	L	1 (43%)
1	5.3,4.3*	0.2,1.2	l_2	-1	VL	0.7625 (33%)
2	3.1,2.1	2.4,1.4	v_1	-0.25	E	0.25 (11%)
3	4.9,5.9	1.6,2.6	v_2	-0.75		
4	3.7,2.7	2.8,3.8	v_3	0.75	$h^2_{\text{w/in location 1}}$	0.84***
			v_4	0.25	$h^2_{\text{w/in location 2}}$	0.77
			vl_{1j}, vl_{4j}	±1.05		
			vl_{2j}, vl_{3j}	±0.65		
			e_{ijk}	+/-.5		

* The two figures separated by a comma denote two independent replications. The order of the figures is of no significance.
** Figures in parentheses give percentages of the total variance.
*** h^2 denotes broad-sense heritability.

Data Set 1a			Estimates of contributions		Variance & heritability estimates	
Location	1	2	m	2.5	V	0.0625 (2.5%)
Variety			l_1	1.2	L	1.44 (58%)
1	5.3,4.3	0.2,1.2	l_2	-1.2	VL	0.7225 (29%)
2	3.1,2.1	2.4,1.4	v_1	0.25	E	0.25 (10%)
			v_2	-0.25	$h^2_{\text{w/in location 1}}$	0.83
			vl_{ij}	±0.85	$h^2_{\text{w/in location 2}}$	0.59
			e_{ijk}	+/-.5	$h^2_{\text{across locations}}$	0.25

Data Set 1b			Estimates of contributions		Variance & heritability estimates	
Location	1		m	3.7	V	1.21 (83%)
Variety			l_1	0	E	0.25 (17%)
1	5.3,4.3					0.7225 (29%)
2	3.1,2.1		v_1	1.1	$h^2_{w/in\ location\ 1}$	0.83
			v_2	-1.1		
			e_{ijk}	+/-.5		

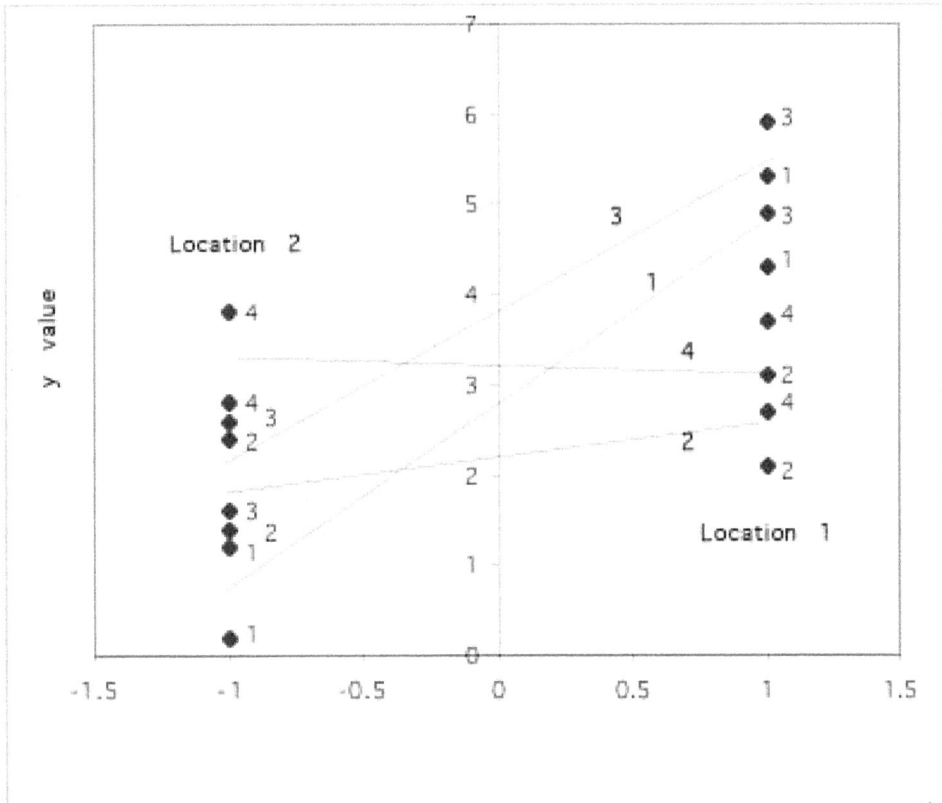

Figure F.1. Data Set 1 from Table F.1. Lines connect the midpoint of the variety in each location. The x-axis is the location contribution, i.e., average over all varieties for that location *minus* the overall mean.

Conditionality can be discounted if the varieties and locations observed are a random subset of all possible variety-location combinations. Careful consideration is needed before asserting that the subset represents the whole population in this fashion. (See as a counterexample, Turkheimer et al. 2003, in which heritability of IQ test scores in young children varies according to the socioeconomic status of the sample of children included in the analysis.)

2. The idea of rerun predictability

There are several ways that heritability and other fractions of the overall variance can be given causal significance without translating these quantities into terms of measurable genetic and environmental factors (Item B, Gap 2, e.). One way is through the idea of rerun predictability.

We have seen that analysis of the variation in a trait can proceed without knowing anything about the underlying genetic and environmental factors, let alone the transmission of factors or dynamics of development that generate the trait (Item D). Suppose we assume that these unknown dynamics include some unsystematic noise and then imagine that the same sets of varieties and locations are observed again where the only change in this *rerun* is noise at the same level, but uncorrelated with, the original. Provided the noise is small and the rerun remains close to the original situation, a good approximation for a wide range of actual, but unknown, dynamics in both situations can be provided by models, like equation 1 in Item D, that add up separate contributions, including ones associated with the variety, with the location, and with their combination.

If we fit such an additive model to the observations and assume that the noise contribution approximates the actual noise, we do not actually have to *conduct* the rerun. Instead, we can *predict* how closely the trait values in each location in the rerun would match the original situation. The closeness of match can be assessed by the correlation of all possible pairs of values in which the first value is from the original situation and the second is given by model 1 with the *location* set to be the same as the original, but the other components not so constrained. Similarly, we can make predictions for how closely the original and rerun trait values would correlate if, for each pair, the original and rerun *variety* is the same (Figure F.2). The first correlation is sometimes called *environmentality*; the second is what we have been calling *heritability* (more precisely, broad-sense heritability across locations; see Box B.1). Both are instances of a general concept we can call *rerun predictability* (Taylor 2006).

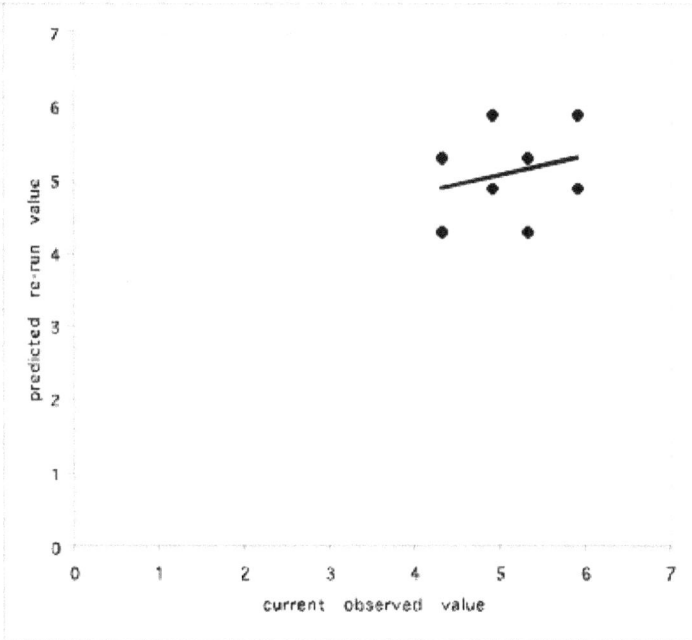

Figure F.2. Varieties 1 to 4 in location 1: Correlation between observed values and predicted rerun values. The lines indicate the correlation between the original and predicted rerun values.

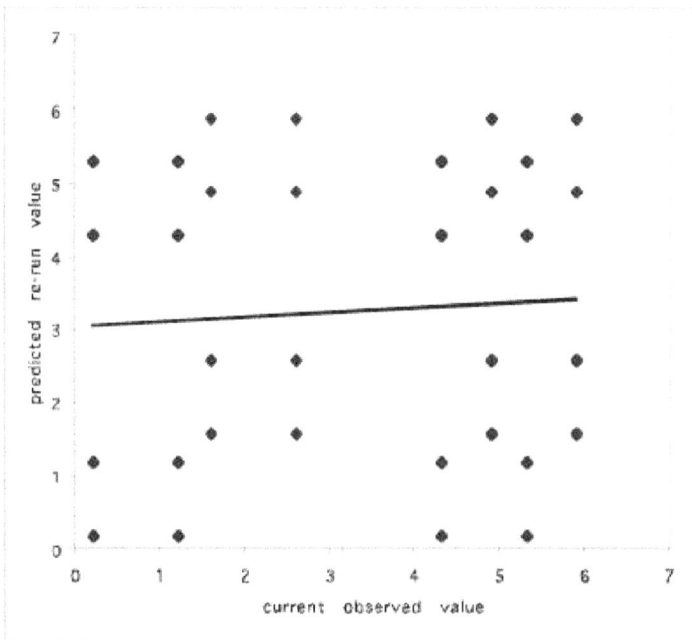

Figure F.3. Varieties 1 to 4 across both locations: Correlation between observed values and predicted rerun values.

Heritability is based on measurements at one point in time, but we can now see how it relates to predictions about the corresponding point in the next generation. The additive model from which heritability is derived can be interpreted dynamically *provided we treat the step to the next generation as a rerun*, that is, if the varieties and locations remain the same and the noise is at the same level as the original generation, but uncorrelated with it. This rerun condition is plausible in agricultural and laboratory settings where researchers have the ability to replicate varieties and locations. (For field trials the rerun condition becomes somewhat less plausible because the variability of weather from season to season may be considerable.) However, in many other situations the next generation cannot be viewed as a rerun situation because the full set of varieties and locations are not (or cannot be) replicated, or because the varieties from different families are mated to produce the offspring generation. In such situations (which include studies of human traits) additive models, and thus heritability, become an unreliable basis for prediction or other actions. (In agricultural research there are ways to address such shortcomings; see Item G).

Equation 1 is by no means the only one used in studies of traits that vary across varieties and locations and in the calculation of heritability. Models that take into account degrees of relationship between varieties or replicates are common and can be fitted to measurements from situations in which varieties are replicated only in one or two locations using either a gene-free formulation (Item D) or path analysis and Structural Equation Modeling. In all cases, however, the calculation of heritability is an estimation of predictability in a rerun situation. (This point is spelled out in the account of path analysis in section 5, below.) Similarly, to calculate the variance of the location contributions (the so-called shared environmental effect) is to calculate rerun predictability.

3. *The calculation of rerun predictability*

For the situation depicted in Figure F.2, j is fixed, which means within-location heritability is the focus. The formula for rerun predictability can be derived as follows:

Correlation between observed and predicted

= Covariance (observed, predicted)/ [Variance (observed)* Variance (predicted)] $^{1/2}$

= Covariance $(m + v_i + l_j + vl_{ij} + e_{ijk}, m + v_{i'} + l_j + vl_{i'j} + e_{i'jk'})$

/ [Variance $(m + v_i + l_j + vl_{ij} + e_{ijk})$ * Variance $(m + v_{i'} + l_j + vl_{i'j} + e_{i'jk'})]^{1/2}$

where j is fixed, but i and k can vary and ' denotes the rerun.

This quantity can be estimated by

Covariance $(v_i + vl_{ij}, v_{i'} + vl_{i'j})$ / [$(V + VL + E)$ * $(V + VL + E)]^{1/2}$

given that the noise (or residual contributions) are uncorrelated and $m + l_j$ is
a constant within any location.

In turn, this can be estimated by

$(V + VL)/ (V + VL + E)$

given that the variety is constrained to be the same in the observed situation
and the rerun (i.e., i = i').

In general, rerun predictability is given by

Variance of the contributions in the additive model that are constrained to
be the same in observed case and the rerun in the defined situation /

Variance of all contributions in the additive model that are not constant in
the defined situation (which includes appropriate noise variance)

(22)

4. Rerun predictability and replication of varieties and locations

Given that heritability and other rerun predictability measures are quantities
that can be derived by fitting measurements to simple, additive models, it
follows that *we have no basis on which to apply or interpret the quantities outside
the circumscribed, rerun situation.* This conclusion makes a certain sense in
agricultural and laboratory research. Recall that a variety refers to a group of
individuals whose relatedness by genealogy can be characterized, such as
offspring from the *repeatable* mating of a certain sire and dam, and it also refers
to a group of individuals whose mix of genetic factors can be *replicated*, as in an
open pollinated plant variety. Locations are the situations or places in which the
varieties are raised. Give or take variability in weather affecting field sites from

season to season, locations can also be replicated. In short, it is feasible in agricultural and laboratory research to make a rerun.

For human research, however, replication is limited or impossible. Yet, when methods are used to analyze variation across varieties and locations ("genotypes" and "environments"), a *thought-experiment* is implied in which replication of varieties and locations is indeed possible. Remember, there is nothing in the definitions of variety and location that presumes that researchers know or can specify the genetic or environmental factors that influence the trait for any variety-location combination. (Of course, researchers may hope eventually to bring measurable genetic or environmental factors into their analyses, but this is another matter; see Gaps 3 and 7 and Table B.1 in Item B and Items H and I.) In the absence of knowledge about the underlying factors in the situation analyzed, it is not possible to make predictions about the trait based on measurable factors in any new situation. With the thought experiment, heritability studies, which proceed from data about the trait alone, can at least say something predictive, something that it not simply a summary of the data set in hand. Without this thought experiment, we could ask just what quantitative geneticists can expect to do with the knowledge from analyzing variation for a trait between human varieties and locations.

5. Path analysis and rerun predictability

Some researchers seem to give causal significance to heritability and other fractions of the overall variance, without translating the quantities into the terms of measurable genetic and environmental factors, when they use the data analysis techniques of path analysis and its generalization as Structural Equation Modeling. The foundations of path analysis, as presented in this section, indicate that when used in heritability studies, it can yield no causal insights beyond those covered by rerun predictability.

Path analysis quantifies the relative contributions—the *path coefficients*—of variables to the variation in a focal variable once a certain network of interrelated variables has been accepted (Lynch and Walsh 1998, 823). The usual starting point for path analysis is an additive regression model that associates the focal variable with several other measured variables. However, it is possible to employ the technique when there are no measured variables except the observed focal variable, as is the case for the observed trait in classical quantitative genetics. This latter form can be arrived at by converting the additive model on which any given ANOVA is based into an additive model of *constructed variables* that take the values of the contributions fitted to the first model. For example, the path model equivalent to equation 1 (page 56) is

$$y_x = \quad m \quad +z_{1x} \quad +z_{2x} \quad +z_{3x} \quad + e_x \tag{23}$$

where

y is the measured trait as before and x denotes the replicate

$z_{1x} = v_i$ if x if a replicate of variety i

$z_{2x} = l_j$ if x if a replicate in location j

$z_{3x} = vl_{ij}$ if x if a replicate of variety i in location j

$e_x = e_{ijk}$ where x is replicate k of variety i in location j, which might be synonymously referred to as z_{4x}

(It should be noted that, unlike uses of path analysis in some other fields, the constructed variables cannot be manipulated through their insertion or removal; Pearl [2000, 135 and 344-5]. See Freedman 2005 for critique of views about manipulable variables.)

Equation 2 becomes the *equation of complete determination* that lies at the heart of path analysis:

$$1 \quad = \quad \Sigma \text{ variance } (z_w) / Y \tag{24}$$

where w denotes the different contributions (1, 2, 3, 4 or v, l, vl, e) in the additive model

and Σ denotes the sum over possible values of the subscript, in this case, w.

Thus far, path analysis is simply an algebraic reformulation of the ANOVA. However, when the same trait is observed in two relatives, their separate path analyses can be linked in one network and the intraclass correlation for the trait between relatives can be calculated (Lynch and Walsh 1998, 826).

$$I_{yy'} = \quad \Sigma \text{ swy } r_{wyy'} \text{ swy'} \tag{25}$$

where y and y' denote the trait for any pair in a class of relatives

I is the intraclass correlation

w denotes the different contributions (v, l, vl, e) in the additive model

swy is shorthand for square root of variance $(z_w)/Y$, i.e., the *path coefficients*

$r_{wyy'}$ is the correlation between the contribution w for relatives y and y'.

Suppose that the contributions and thus path coefficients apply to both relatives and that the noise contributions are uncorrelated—in other words, if the rerun conditions apply (section 2). Equation 25 simplifies to:

$$I_{yy'} = \qquad \Sigma \ r_{wyy'} \ \text{variance} \ (z_w)/Y \qquad\qquad (26)$$

with symbols as given for equations 24 and 25

When this last equation is applied to the network defined by Figure F.4, in which offspring need not be raised in the same location as parents, the predicted intraclass correlation between parent and offspring over all locations is V/Y, the broad-sense heritability.

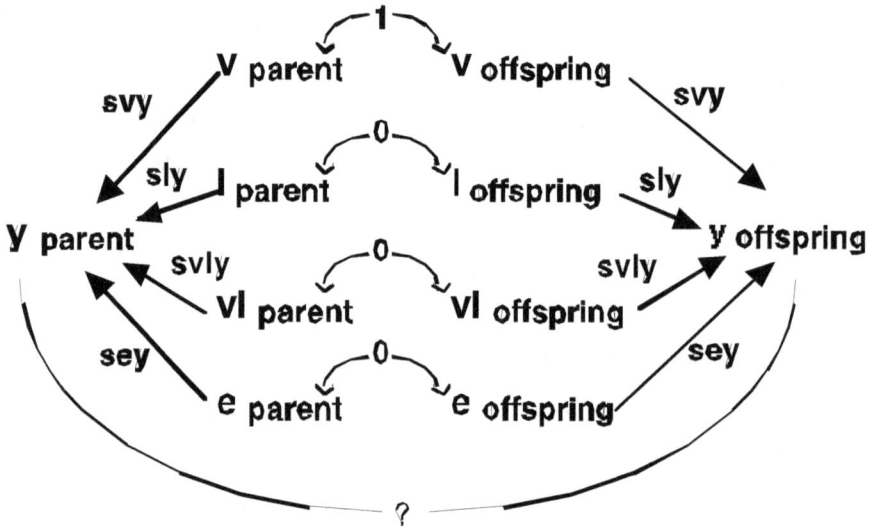

Figure F.4 Path diagram for Parent-Offspring relationship in which offspring need not be raised in the same location as parents. svy is shorthand for square root of V/Y; sly for square root of L/Y; etc.

Networks of interrelated variables more complicated than Figure F.4 can be analyzed. Typically, these incorporate diploidy and biparental inheritance, degrees of relatedness of different varieties (not only parent-offspring pairs), and correlations between replicates (e.g., when plots are assigned non-randomly within a single location). (In the analysis of human traits, non-random replication is usually attributed to some unspecified environmental factors being shared by siblings in a family or differing between siblings.) In other respects, however, networks in heritability studies are simpler than Figure F.4. Typically, networks omit VL and reflect Assumption 2 from Item E, that is, all other things being equal, similarity in traits for relatives is proportional to the fraction shared by the relatives of all the genes that vary in the population (e.g., γ = .5).

Putting aside the complications and simplifications, path analyses in heritability studies have three features shared with ANOVA and rerun predictability:

a) The assumption, already mentioned, that contributions in the additive model and path coefficients are constant across generations.

b) The analysis is based on observed traits and does not require reference to measurable genetic factors that are transmitted from parent to offspring.

c) The contributions and path coefficients are conditional on the particular set of genetically defined varieties and locations observed (from which contributions or path coefficients are estimated). (See section 1 above.)

These features mean that the contributions and path coefficients have no connotations of causality beyond rerun predictability.

6. Historical puzzles: Translation from agriculture and laboratory breeding to human genetic analysis

If, for studies of human variation and heredity, it seems of limited interest to consider rerun predictability in the thought-experiment in which replication of varieties and locations is possible, we might wonder, picking up on Puzzle 3 in Items A and C, why heritability studies have been carried over from agriculture and laboratory breeding to human genetic analysis.

Heritability estimation was first used in selective breeding in agricultural and laboratory settings (Fisher 1918, Wright 1920, Lush 1945). In this context, as noted before, researchers have the ability to replicate varieties and locations. Indeed, when agricultural researchers compare varieties and make recommendations to farmers and when they select between varieties for the next round of evaluation trials, they do so on the assumption that the environmental factors will remain more or less unchanged. For observations of human traits, however, such replicability of varieties and environmental factors is not possible (requiring the thought-experiment discussed in section 4). How then were such restrictive conditions discounted or forgotten in the translation of heritability estimation from selective breeding to human genetics?

When Wright (1920) presented his original formulation of heritability estimation, he used the notation E to refer to "environmental factors that are common to litter mates" of guinea pigs that he bred. To translate heritability estimates into predictions of future changes under selective breeding, these "factors" had to remain constant from one generation to the next. (No measurable factors were involved in the analysis; what Wright meant was what this book calls the contributions, as estimated through the ANOVA or path analysis, had to remain constant.) "E" is now used, however, to denote environmental factors without reference to Wright's restricted conditions. One part of an historical investigation would be to trace Wright's notation from its origin through its adoption in human genetics (see Burks 1928; Lush 1947), where it remains commonplace to discuss the relative influence of G (genes) and E (environment) in accounting for the variation between individuals and groups. (Other historical investigations might consider the influence in the other direction [D. Paul, pers. comm.], the separation of heritability from the context of selective breeding [S. Downes, pers. comm.], or even the grounding of Wright's shifting balance theory of evolution in the search for ideal approaches to animal breeding; see, e.g., Lush 1945, p. 433–435.)

The historical question concerns the forgetting of the restrictive conditions of selective breeding. The question could be extended into a critical revisiting of the long and politically charged history of scientific and policy debates about the heritability of IQ test scores (and other human traits), and about genetic explanations of the differences between the mean scores for racial groups. (For entry points, see Puzzle 1 in Item A, Lewontin 1970a,b, Jensen 1970, Schiff and Lewontin 1986, Jencks and Phillips' 1998 review of research on the black-white test score gap, Nisbett et al. 2012, Parens and Chapman's 2006 overview of past and potential contributions of human behavioral genetics to discussions of social importance well beyond IQ tests, Sesardic's 2005 critique of critics of human behavioral genetics.) Such critical revisiting, as well as further inquiry into the historical puzzles above, lies beyond this book's scope.

*   *   *   *   *

Once heritability and other fractions of the variation in the trait are subsumed under the general category of rerun predictability, their causal significance is very circumscribed—for heritability studies as a whole, not only for analysis of human data. The next item teases out the conditions in which results based on additive models and partitioning of variation might be ascribed some meaning beyond the rerun situation.

Item G. Underlying heterogeneity, the dynamics of development, and the bases we have for action

In which a richer notion of causality, one that draws on the actions possible or proposed on the basis of what is known, is applied to heritability studies in light of the possibility of heterogeneity in the genetic and environmental factors underlying the trait.

The Short Critique of Part I concluded that methods of heritability studies cannot show anything clear and useful about genetic and environmental influences. In agriculture and laboratory breeding, however, the practice of selection means that these fields can make progress without knowing about the dynamics of development of the trait, the specific genetic and environmental factors involved, or whether those underlying factors are heterogeneous. By clarifying when and why progress in breeding is possible, this item characterizes the scope and limitations of heritability studies more fully, positioning those studies as one option among others for managing a two-way knowledge-action relationship. In brief, the relationship is one in which, even if we try to address our limited knowledge about the dynamics of some phenomenon through further research, whether that research is necessary depends also on the actions possible or proposed on the basis of what is known—or unknown (Taylor 2005, pp. 93ff).

Agricultural and laboratory research allows for the pursuit of options not feasible in human heritability studies. In order to illuminate what can and cannot be derived in the less controlled and thus less informative circumstances of research on similarity among humans, this item keeps the focus wide so as to include agricultural research (albeit somewhat idealized) in which a set of varieties is raised in multiple replicates over a wide range of locations. Agricultural research, with its control of the replication of varieties and locations, seems well suited to clarifying the kinds of realizable interventions that are involved when making inferences about causality from observations of traits. This task fits with philosophy and social science on the causation-intervention relationship (Pearl 2000, Woodward 2003).

In the context of examining causation-intervention relationships, the responses to the possibility of underlying heterogeneity discussed in the Short Critique (see Puzzle 2 in Items A and C and Gap 6 in Item B) are reconceptualized here as five conceptually distinct approaches. These approaches allow us to consider, to different degrees, the reliability of actions taken (or assumed) by researchers in relation to the degree of knowledge of the dynamics through which the trait develops. (Heritability studies will be seen to fit under the fifth approach.) Three of the approaches ignore the possibility of underlying heterogeneity; the other two attempt to reduce it by different means. The preamble below is intended to help readers visualize the meaning of underlying heterogeneity used in this book.

0. Visualizing underlying heterogeneity

When similarity between a set of close relatives (such as twin pairs) is associated with similarity of (yet-to-be-identified and measured) genes or genetic factors, those factors are not necessarily the same from one set of relatives to the next. This possibility, that of underlying heterogeneity, was introduced in Part I's Short Critique of heritability studies. Let me work through this idea using some schematic diagrams and provide a more technical note distinguishing it from some other senses of the term *heterogeneity*.

The bell curve at the top of Figure G.1 depicts the relative frequency of a trait, say, height (which increases from left to right), in a group of individuals (each denoted by *). During the growth of these individuals, their height has been influenced by a set of environmental and genetic factors. Here only one of each kind (ef and gf) is depicted; the lines point to these underlying factors for a sample of the individuals. If individuals who have similar heights do not necessarily share similar underlying factors, the lines will cross, as is the case in Figure G.1, which provides a simple depiction of the idea of underlying heterogeneity. The idea makes intuitive sense for height—we are familiar with some people being long-legged while some are long in the trunk so the same height for two people may be made up of quite different body types. Similarly, the growth trajectories through adolescence can be quite disparate.

Figure G.2 depicts the frequency of a trait, such as height, for two groups of people raised in two situations, a and b (which may or may not be the same). (Here, four underlying factors are included and are linked so as to indicate that the influence of a genetic factor on the dynamics of development is dependent on the environmental factors, and vice versa.) If there is heterogeneity of under-

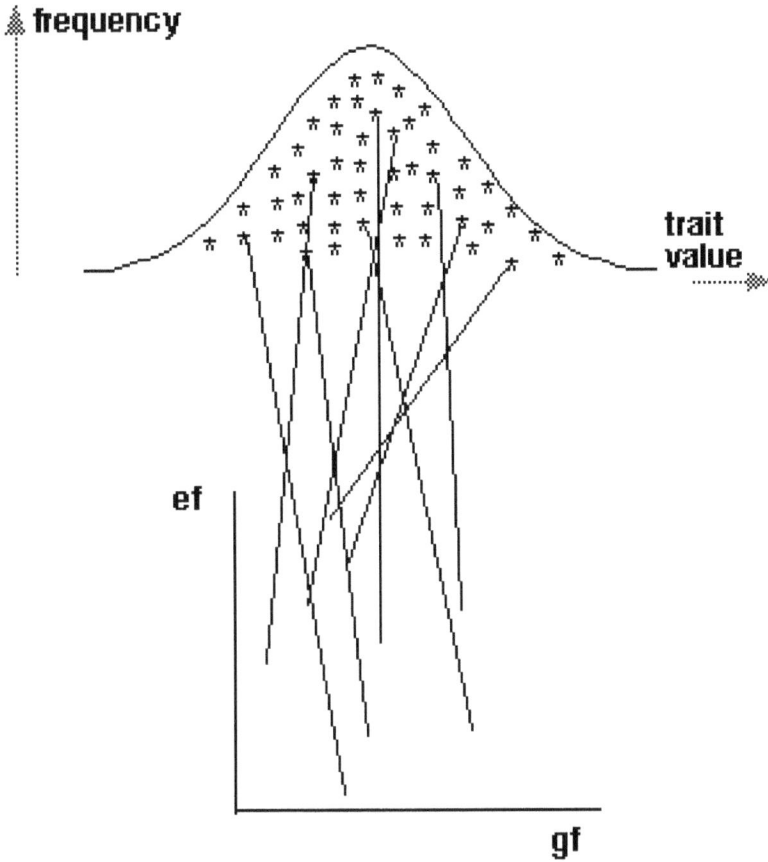

Figure G.1. Values of a hypothetical trait for a population, showing one underlying genetic factor and one environmental factor (ef and gf) connected to the values for a sample of individuals. The crossing of the connector lines indicates schematically that the underlying factors are heterogeneous for this trait.

lying factors, there are many subsets of group A who share more of the genetic and environmental factors that influenced the development of their height with some subsets of the group B than they do with the other individuals in group A.

Such heterogeneity causes problems for the conventional mode of explaining the difference between individuals in the two groups, a mode that involves two steps: *first* we account for the difference between the average values for the groups, *then* for the difference of the individual from their group's average. In the conventional mode we would say "Euro-American females are, on average, taller than Asian-American females" or "Watutsis are taller than Pygmies." In contrast to such two-step accounting we could take the spread within both of

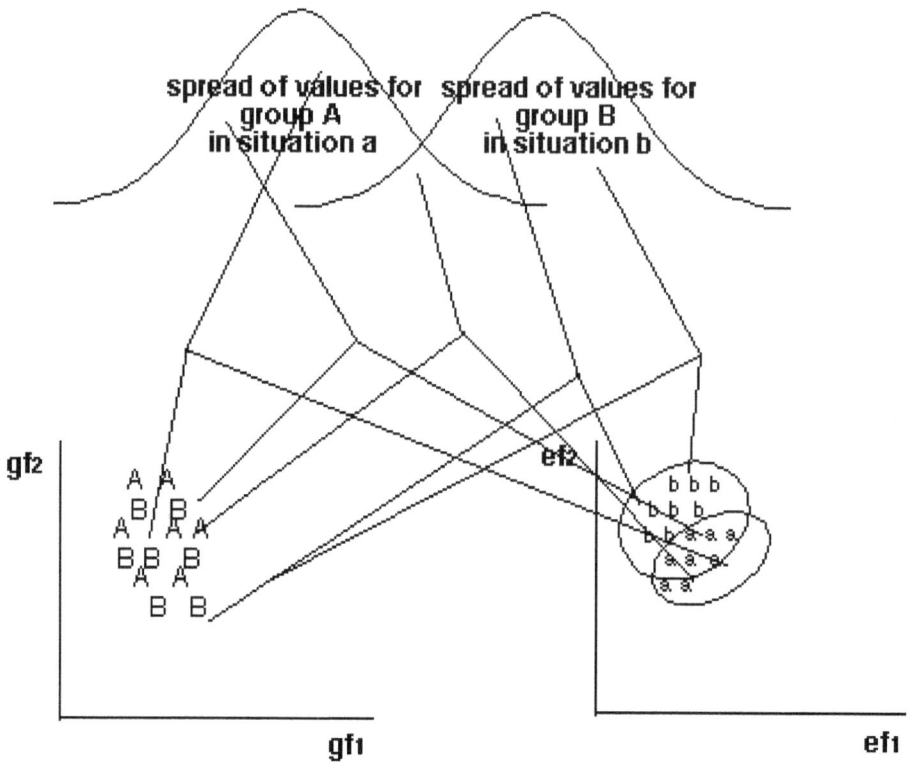

Figure G.2. Trait values in two groups (A and B) connected to the underlying genetic and environmental factors (ef and gf) for a small sample of individuals. The overlap of the two groups' genetic factors and environmental factors indicates schematically that the factors underlying the trait are heterogeneous.

the two groups as our starting point and attempt to expose the heterogeneous combinations of factors that influence the development of the trait for the range of individuals (Taylor 2006b).

The picture of underlying heterogeneity does not rule out the possibility that within each group certain genetic and environmental factors are common to the development of all individuals. (Common factors would mean that differences between those factors are shared by each pair of individuals from the two groups.) However, explanation would not *begin* by assuming that is the case, let alone assuming that factors shared within a group were the dominant ones in accounting for differences between individuals. Such patterns would have to be demonstrated, not assumed.

An equivalent contrast applies to heritability studies, for which the group is the variety. For example, in the case of twin pairs in human studies, a two-step approach would account for the differences among the averages for the twin pairs (invoking, say, gradients of genetic or environmental factors), and would *then* interpret the deviation from those averages. Accounting for heterogeneity (as depicted in Figure A.2) is the subject of the sections below.

Figures G.1 and G.2, along with A.2, should make clear that the heterogeneity of underlying genetic and environmental factors encompasses more than *genetic* heterogeneity, in which there are either: a) a multiplicity of mutations within some gene and a spectrum of corresponding values for a trait (or "phenotype"), each of which varies little within the typical range of locations (i.e., *allelic* heterogeneity); or b) the trait exists if any one of a range of loci has the atypical form (i.e., *locus* heterogeneity). These two cases of underlying heterogeneity for a trait might disappear if the trait could be resolved into separate traits, each one corresponding to a different allele or locus. Some other cases might disappear if the trait is revealed to be a composite of separable traits (analogous to the time taken to complete a triathlon being made up of the time taken to complete the running, swimming, and bicycling events). There is no reason, however, to assume that all cases of underlying heterogeneity are mixtures or composites of separable traits.

Underlying heterogeneity also differs from *statistical* heterogeneity, in which two sub-populations differ in their statistical moments (mean, variance, etc.) or, if multiple traits are measured, in their variance-covariance patterns. In these cases, the population as a whole might be better treated as a mixture of distinct populations. Yet, subdivision into two or more populations does not automatically eliminate underlying heterogeneity in each population separately.

More will be said about underlying heterogeneity in Part III, but the schematic diagrams and the note on terminology in this preamble should suffice to set the scene for the five following approaches.

I. Ignore underlying heterogeneity, then compensate if problems arise

What can we do on the basis of knowing that heritability is high for a trait when the factors underlying that trait might be heterogeneous? We know, by the very definition of heritability, that differences among the average values for the varieties make up much of the total variation. So a breeder might mate (or cross) individuals with the desired values for the trait, expecting that this will lead to offspring with similar, desired values (and thus to improvement in the overall average compared with the previous generation). However, underlying heterogeneity creates the possibility that the offspring of individuals from separate families raised in the locations of their parents (in the example in Puzzle 2, AabbCcDDEe subject either to the factors FghiJ or FgHiJ) depart from this expectation. A similar concern arises when there is selection without mating, which is possible for species in which varieties can be replicated and raised again (using, e.g., inbred lines, stored seeds, or plant clippings). High heritability might lead us to expect that choosing the best varieties from the first round and raising only this subset for a second round will lead to improvement in the overall average. This outcome must be the case if the varieties are raised in the same place the second time and the weather and other not-always-obvious conditions repeat themselves. However, if the locations are *not* the same, then, before we could be confident in that expectation, we would need to know more about the dynamics of development (or, at least, the degree to which the ranking of varieties changes from location to location) and to rule out the possibility of underlying heterogeneity.

Notwithstanding these limitations, agricultural and laboratory breeders can proceed as if there were no underlying heterogeneity, assess whether the results meet their expectations, and try to compensate when the results do not. In the example above, if it turns out that there is underlying heterogeneity and that the reassortment of genes from parents leads to some far-from-expected offspring, breeders can discard those offspring, selecting only the offspring that do have the desired values. This approach may be summarized as "Ignore, then compensate." It is not applicable, however, to the study of human traits, where selective mating and discarding of defectives are not acceptable.

What bases do breeders have for the expectations that underwrite their actions? Given that they can compensate for the shortcomings of their decisions, they need not know much about the transmission of genetic factors and dynamics of development. In the example above, breeders can make improvements in the population *from one generation to the next* even though their expectations are based on heritability, which is a summary of measurements made *at one point of time* for a specified set of varieties and location. They can, moreover, make improvements even if the models on which their expectations are based do not capture the dynamics of the development of the trait or are sensitive to the key assumptions of Item E (i.e., about idealized

Mendelian models, trait similarity proportional to fraction of genes shared, and discounting of variety-location interaction). (Breeders' expectations are based on calculations that link degrees of relatedness of parents, heritability, and simple models of theoretical genetic and environmental factors that contribute to the trait; Falconer and Mackay 1996, Lynch and Walsh 1998.) Breeders can ignore the lack of realism of such models as well as the possibility of underlying heterogeneity because they can continue selective breeding for more generations to compensate for any less-than-expected progress over any one generation.

In the study of human traits, selective breeding is not acceptable, so it is not relevant to link heritability to expected progress under selection using formulas based on simple models, or to compensate if problems arise.

2. Reduce the possibility of underlying heterogeneity by grouping varieties that are similar in responses across locations

In agricultural trials, where a number of varieties or animals or plants can be raised or grown in multiple replicates in many locations, varieties can be grouped by similarity in responses across all locations using techniques of cluster analysis (Byth et al. 1976). (Similarly, locations can be grouped by similarity in responses elicited from varieties grown across those locations.) Varieties in any resulting group tend to be above average for a location in the same locations and below average in the same location (Figure G.3). The wider the range of locations in the measurements on which the grouping is based, the more likely it is that the ups and downs shared by varieties in a group are produced by the same conjunctions of measurable factors. This likelihood of homogeneity within the group gives us more license to discount the possibility of underlying heterogeneity. Thus Approach 2 may be called "Reduce by grouping."

If the underlying factors are assumed to be homogeneous within each of the groups, we can hypothesize about the group averages—about what factors in the locations elicited basically the same response from varieties in a particular variety group that distinguishes them from other groups. (It should be noted that knowledge from sources other than the data analysis is always needed to help us generate any hypotheses about genetic and environmental factors.) For example, imagine a group of plant varieties that originated from particular parental or ancestral stock that is more susceptible to plant rusts (a form of parasitic fungi), and that these varieties had a poor yield in locations where rainfall occurred in

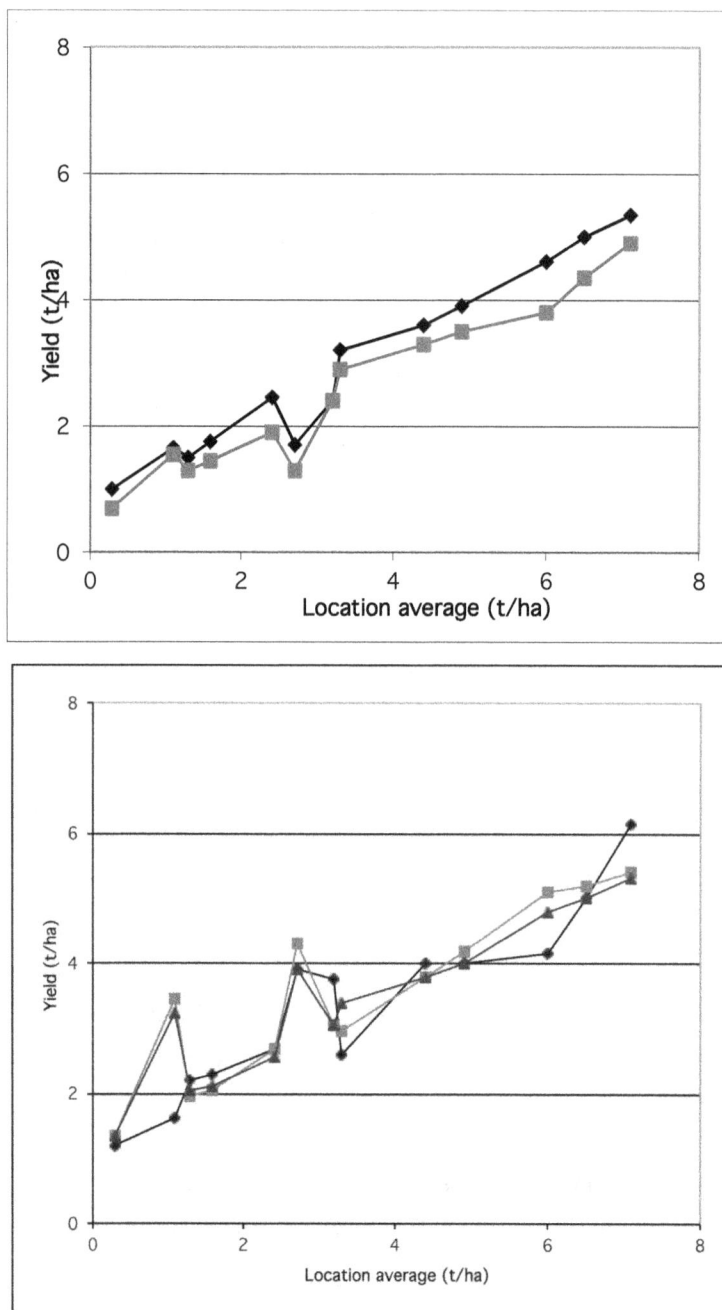

Figure G.3. Yields for 5 groups of wheat varieties grown in 13 groups of locations (from Byth et al. 1976). The x-axis is the average over all varieties for that location. The individual varieties (not shown) were clustered into these five groups by similarity of response across locations. These groups were then clustered into two groups as shown in the two plots.

concentrated periods on poorly drained soils. The obvious hypothesis about genetic factors modulated by environmental factors is that these varieties share genes from the parental stock that are related to rust susceptibility and this susceptibility is evident in the measurements of yield in locations where the rainfall pattern enhances rusts. Through additional research to compare the variety and ancestral genomes, it may be possible to identify specific sets of genes that are shared, to investigate whether and how each one contributes to rust susceptibility, and to use that knowledge in planting recommendations for locations like those observed in the trial and for subsequent research that might extend beyond the observed varieties and locations. In general, any hypotheses we generate need to be validated before making proposals for action. If hypotheses are not forthcoming, or if they fail to be validated, it is possible to shift back to Approach 1, this time around making use of one of the groups, not the whole data set, in decisions about selection and breeding.

A brief digression: What role does heritability play in research that groups varieties so as to reduce underlying heterogeneity? As it turns out, very little. Clustering ensures that the variation between the means for variety groups is much higher than the average variation between variety means within groups. The low within-group variation allows the selective breeder to select from the variety group without being very concerned about whether any one variety within that group is the best across locations. In other words, heritability within the variety group is not very important to the breeder. Between-group heritability also lacks importance—even if variation between variety group means is smaller than variation between location means, researchers can still hypothesize about the group averages.

Back to the thread of grouping varieties that are similar in responses across locations: It becomes more difficult to distinguish groups of varieties by similarity of responses across locations when varieties are observed in only a few locations or when the locations are not the same from one variety to the next. Approach 2 becomes infeasible when analyzing measurements from studies of human twins because such studies have only two replicates (twins) in one or at most two locations (families). (This also means that grouping varieties in this way is not a direction by which research on human variation can bridge or circumvent Gaps 3 and 6 [Item B], which concern the difficulty of translation from components of variation to hypotheses about measurable factors.)

3. Reduce the possibility of underlying heterogeneity by restricting the range of varieties or locations

In agricultural research, we can also reduce the possibility of underlying heterogeneity by restricting the range of locations in which a variety is raised or grown. We can also control environmental conditions, such as, for animals, the regimes of feeding and husbandry or, for plants, the application of fertilizer and irrigated water. Agricultural breeders can also produce inbred lines and thereby eliminate the heterogeneity of genetic factors that exists within outbred varieties. However, to consider taking some action on the basis of research conducted under restrictive conditions is to presume that the restrictive conditions can be replicated. This is most apparent when plant breeders recommend varieties to be grown only in defined regions and under prescribed techniques of cultivation, or when animal breeders specify the optimal feeding regime for each variety. Thus Approach 3 might be called "Reduce by restricting and replicating."

In the study of human traits, it is not feasible to control the full range of relevant environmental conditions or to breed for genetic uniformity. It may be possible, however, to restrict the locations included in a human study (e.g., to exclude families of low socioeconomic status; Turkheimer et al. 2003) and, even without identifying the underlying environmental factors, to replicate these restrictions in subsequent research. These restrictions need to be brought to light to prevent any unqualified interpretations or policy proposals.

4. Ignore underlying heterogeneity, then check

We can try to expose underlying factors using some method that assumes that the factors are *not* heterogeneous and later check whether the results hold up under further scrutiny. If the method helps us in some way to identify genetic factors, we then investigate whether these factors have their effect consistently across locations—we investigate whether, every time the trait is at, say, level x, all the same genetic factors are present. In other words, we check whether or not the method is confounded by heterogeneity for this trait—in short, "Ignore, then check."

The simplest example of a heterogeneity-ignoring method would be to assume that varieties that are more similar in their averages for the trait over all locations in which they are measured might be so because they share more genes or genetic factors. (In human studies we might restrict our attention to single sets of close relatives, e.g., Brunner et al. 1993.) We would then search for those factors. (As in Approach 2, knowledge from sources other than the measurements themselves is needed to suggest what the specific factors might be.) If it turns out that a genetic factor we had identified does *not* have a consistent effect across locations, this is not the end of the road. We might shift to another method that helps us to identify for specific variety-location combinations which environmental factors are modulating those genetic factors. (Such a method may well combine the method for identifying genetic factors with an equivalent one for identifying environmental factors for locations that are more similar in their averages for the trait over all varieties.) If so, we can investigate whether these factors have their effect consistently across variety-location combinations.

If it turns out that the underlying genetic and environmental factors do not have their effect consistently across variety-location combinations, several options still remain. We could shift back to Approach 2, restrict our attention to one of the homogeneous groups, and re-apply whatever methods we were using to help expose underlying factors. Alternatively, we could reduce the likelihood of underlying heterogeneity by raising the varieties in controlled environmental conditions (see Approach 3). Indeed, the increasingly popular technique of mapping quantitative trait loci (QTL)—regions of the genome containing genetic factors that influence a continuously variable trait—has had most success in animal and plant varieties that can be replicated and raised in controlled conditions; reliable QTL results for human populations are few (Majumder and Ghosh 2005; but see The Wellcome Trust Case Control Consortium 2007). A final option is to leave the chosen case behind and look for other cases in which the results are not confounded by underlying heterogeneity. Of course, if finding such cases entails sieving through many others, the few that are *not* confounded should be seen as *special* cases. They should certainly not be invoked to suggest that genetic determination of traits is the norm.

5. Ignore underlying heterogeneity when fitting data to simple models of theoretical underlying factors, but acknowledge limitations

If we lack hypotheses about the underlying factors, we can always make simplistic assumptions about the influences on the dynamics of development, build a model based on those assumptions, fit the model to the measurements, and assess whether the model fits well—or fits better than alternatives. The risk with such an approach is that we start to think and act as if a model with its assumptions refers to real relations among real entities *even though we have no evidence for the assumptions independent of the models' fit to the measurements* (Taylor 2005, pp. 35ff; see Item D and E for gene-free alternatives to the standard models used in heritability studies). Prudent use of this approach acknowledges its limitations and refrains from using the model to hypothesize about the actual dynamics. We should only expect predictions made using the model to be reliable when the new situation is a *rerun*, that is, if it stays close to the one from which the model was fitted. In studies of traits varying across varieties and locations, *close* entails the set of varieties and locations remaining the same in the new situation as in the original (Item F). In short, Approach 5 is "Ignore, but acknowledge limitations."

How might we overcome the limitations of this approach for addressing the possibility of underlying heterogeneity—an approach that gives us no warrant to apply or interpret the quantities outside the circumscribed rerun situation? We should, I contend, shift back to one of the other four approaches, in particular, for the study of human traits, to Approach 4 ("Ignore heterogeneity, but check"). However, many moves are made to try to work around the limitations of the rerun conditions. These moves and their shortcomings are reviewed in section 7 below and in Item H.

6. A map of paths, in theory and in practice

The five sections above have sought to identify the nexus between underlying heterogeneity, the dynamics of development, and the bases we have for action (Figure G.4). For example, when researchers are able to compensate for any actions taken on the basis of loose ideas about dynamics, they can afford to ignore the possibility of underlying heterogeneity (Approach 1). If they can restrict the range of varieties or locations, they can reduce the possibility of underlying heterogeneity, but any knowledge they gain about the dynamics presumes that they can sustain the restricted conditions (Approach 3). And so on. The key issue becomes not to pin down whether underlying heterogeneity is indeed the case, but to appraise how any given method addresses or ignores that possibility *in relation to the actions taken (or assumed) by researchers and the degree of knowledge they seek regarding the dynamics of development.* (Such critical appraisal can be applied to other methods for integrating genetic and environ-

Underlying Factors *Heterogeneous*	Dynamics of Development *Unknown*	Bases for Action *Unreliable*
	5. rerun approximation using additive model	[extensions beyond rerun not warranted]
Ignore	1. expectations linked to loose ideas about dynamics ⟶	& compensate for shortcomings
	4. hypothesize about factors* & check whether confounded by heterogeneity	[do not take rare unconfounded cases as norm]
Reduce, 2. by grouping	hypothesize about factors* applying to groups	validate hypotheses before making proposals for action
3. by control	hypothesize about factors*	[actions presume restrictive conditions]
Homogeneous	*Known*	*Reliable*

Figure G.4. Five approaches summarized with respect to the possibility of underlying heterogeneity, knowledge about the dynamics of development, and our bases for action. The order from top to bottom reflects the shift from heterogeneity to homogeneity in underlying factors and the degree of knowledge about the dynamics through which the trait develops. Approaches 1 and 2 do not apply to studies of similarity among human relatives and Approach 3 applies only in a limited fashion (see text for elaboration). (* denotes knowledge has to be drawn from sources other than the data analysis.)

mental factors in the social and the life sciences; see Part III.)

These five approaches can be viewed in relation to a longstanding tension within agricultural research between, broadly speaking, *breeders* and *physiologists*. Breeders seek improvement through selection of varieties combined with plant or animal husbandry appropriate to those varieties. Physiologists focus on determining and manipulating the specific genetic and environmental factors underlying the development of the trait in question. (In this era of genomics, breeders may also be physiologists, but let me continue distinguishing the two ideal types.) Breeders are not uninterested in the underlying factors. They make hypotheses about such factors based on variety-location trials as well as sources other than the data analysis, then use these hypotheses to plan the next set of varieties and locations on which to collect data (see Approaches 2–4). Physiologists make much less use of variety-location trials to generate hypotheses; instead they focus on experiments under controlled conditions. Since the advent of DNA technologies, their experiments have included modification of specific genetic factors. The question that arises is whether the factors that are important under the controlled conditions also apply more broadly (Approach 4)? If not, the controlled conditions have to be prescribed and replicated in actual plant or animal husbandry. The breeder is not so constrained. Even in the absence of realistic models of the underlying factors and even if those factors are heterogeneous, it is sometimes possible that progress through selection and mating can be made (Approach 1).

Where does Approach 5 fit in? For breeders, high estimates of heritability can guide the choice of varieties to select and mate under Approach 1. Breeders can compensate for any less-than-expected progress from one generation to the next by continuing selective breeding for more generations. Breeders may be quite content not only with unreliable estimates made using the standard assumptions (Item E) but also with the application of results from heritability studies beyond the rerun conditions (Item F). Physiologists, on the other hand, want to design experiments around genetic and environmental factors that are likely to be important influences on the trait being studied. In that spirit, although data analysis of variety-location trials is not their primary mode of research, it may be helpful for them to pay attention to the problem points in the standard derivation of quantitative genetics (Item E.4.4) and recognize that heritability studies give no warrant to apply or interpret the quantities outside the circumscribed rerun situation (Item F).

How does this breeder-physiologist tension play out for human heritability studies? After all, controlled mating and selection is not possible, nor are trials based on each of a set of varieties raised in each of a set of locations. Approaches 1 and 2 do not, therefore, apply to studies of similarity among human relatives.

Approach 3 applies only in a limited fashion. (One place where the breeder perspective comes up is when researchers make inferences or speculations about *past* patterns of mating and differential reproduction in relation to *current* traits they identify in social behavior, e.g., Betzig et al. 1988. In such work, however, past variation and heritability of the traits in question are simply assumed; no data analysis is performed.) On the physiologist side, there are obviously serious limits to the use of experiments to determine and manipulate the specific genetic and environmental factors underlying the development of human traits. These limits still, however, leave some room for Approach 4, in which we try to expose underlying factors using some method that assumes that the factors are *not* heterogeneous and then check whether the results hold up under further scrutiny. Yet, whatever method is used to expose the underlying factors, let us ask how important heritability studies are to it.

Consider the generic method used as an example when introducing Approach 4, one in which it is assumed that varieties that are more similar in their averages for the trait over all locations in which they are measured might be so because they share more genetic factors. Then we search for those factors. At first the assumption might seem plausible given the genes involved in the standard derivation of heritability studies, but these are theoretical genes (Item E). In the absence of evidence for these theoretical genes, and given that a gene-free derivation of heritability studies is possible (Items D and E), the plausibility of the assumption becomes less obvious. The grounds for it diminish further for human studies, where the "averages for the trait over all locations in which they are measured" are averages over, at the very most, a few locations. Then, even if the assumption turns out to hold for some human traits, it would remain to be seen how much, in practice, that fact helped in our search for the genetic factors underlying the trait. (The same is true for the equivalent assumption for identifying environmental factors for locations that are more similar in their averages.)

The preceding picture is one in which the relevance of heritability studies to the physiologist side of research is uncertain. It is not surprising that lines of research have developed that address specific, measurable genetic and environmental factors for human traits without making reference to a trait's heritability or the other fractions of the total variance. Such research is the topic of Item I. However, to fully appreciate the scope and limitations of heritability studies, it is valuable not only to draw the contrasts among the five approaches above, but also to consider how researchers proceed when they have *not* accepted the limitations of Approach 5. The idea that heritability studies can help expose or, at least, provide insight about the underlying factors is teased out in section 7, below. The section will set the scene for Item H, which examines specific attempts in human studies to combine data on variation among varieties and locations with data on measurable genetic or environmental factors. Attempts, in other words, to try to hold on to both the breeder and the physiologist sides of the five approaches.

7. Heritability studies and attempting to get around the limitations of Approach 5

Several moves can be made that try to ascribe meaning to heritability studies beyond the rerun situation. This section reviews the moves; the shortcomings, in each case, are noted.

a) Accept the model as a *reasonable approximation*. It might seem that the additive model could be a reasonable approximation to the factors influencing the dynamics of development of traits outside the strict rerun conditions. Indeed, extensions of additive models beyond the rerun situation provide the bases for the expectations used in selection and breeding (see the example under Approach 1, and Lynch and Walsh 1998). To know how robust the approximation is, however, or how quickly it breaks down, we would have to know the actual dynamics—or, at least, to know something about the underlying genetic and environmental factors. Yet, recall that we entertain the simple models of Approach 5 only because we lack hypotheses about these underlying factors.

b) View the model's components as a *synthesis of many factors of small effect*. We can link some traits, such as extra digits, to mutations at a single genetic locus, so it seems plausible to assume that other traits could be subject to the combined influence of genes at a number of loci. In this sense, it might seem that additive models could have a direct relationship to real (but not-yet-identified) genetic and environmental factors. Thus the contributions for varieties could be depicted as a synthesis of the small effects of many different unknown genetic factors (i.e., of pairs of alleles at a large number of loci) (Lynch and Walsh 1998, pp. 81ff). But why adopt this depiction? Under Approach 5 we have no evidence for the assumptions independent of the models' fit to the measurements; indeed, the formulas for heritability can be derived without reference to theoretical genetic factors (Item D). Moreover, models in which the contributions of varieties are assumed to synthesize many genetic factors, each having a small influence, make no allowance for the heterogeneity of the genetic or environmental factors underlying the observed traits (Puzzle 2). The synthesis of many small factors assumption makes it difficult, therefore, to conceptualize or investigate the possibility of heterogeneity (through, for example, Approaches 1–4). The converse is not the case: to allow for the possibility of heterogeneity does not rule out the existence of traits for which the many small underlying genetic factors turn out to be the same for all individuals who show the same value for the trait.

c) *Transfer values into a new situation*. It might seem plausible to extrapolate contributions derived from one set of varieties and locations (where contributions are defined by the specific additive model being used, e.g., equation 1). That is, we would use the same values when any of those varieties and locations are included in another set of measurements—or, at least, would take the values as a first approximation. The problem with this line of thinking is that the contributions for variety or locations derived from fitting an additive model are not properties of the separate varieties and locations. The contribution for each variety in equation 1 (see Item D), for example, can be estimated by the average of the trait over all locations and replicates of that variety minus the same average over all varieties. This means that the value is conditional on the specific set of locations in which it is raised and the specific set of varieties with which it is compared (Item F.1). (This is the case for path analysis [see Item F.5] and Structural Equation Modeling, even though those methods can be applied without explicit estimation of variety and location contributions in the relevant additive model.) Another way of expressing this conditionality is to note that, if the variety and location contributions are labeled as genetic and environmental factors (or "effects"), they are not factors that can be observed and measured. They are factors that can only be derived by fitting a given model to the particular set of observations of the traits for the original set of varieties and locations.

d) Use the model to *inform hypothesis generation*. Suppose that we discounted this last dependency. The contributions for variety, location, or variety-location combinations in equation 1 might then seem to provide some information to help in generating hypotheses about underlying genetic and environmental factors. Recall the example given for Approach 4 ("ignore heterogeneity, but check"), namely, a method of hypothesis generation that assumes that varieties similar in their averages for the trait over all locations may be so because they share more genetic factors than they do with dissimilar varieties. This method is *equivalent* to assuming that, in the rerun situation, varieties that have similar contribution values in the given additive model share more genetic factors than varieties with dissimilar values.

A general problem in following this line of thinking is that the same value for a trait may be the outcome of different combinations of genetic and environmental factors in different individuals. Approach 2 reduces this possibility of underlying heterogeneity, which makes it a more reliable basis for hypothesis generation. Another problem, specific to studies of human traits, is that, under the standard assumptions, the estimates do not actually separate the contributions for variety and variety-location combinations (Item E; see also Item H.4).

e) Take *relative size* into account. Despite the preceding shortcomings, we might try to use the size of heritability and the location component of variation ("shared environmental effect") to inform our choice of cases in which to search further for genetic and environmental factors underlying traits (i.e., path b. under Gaps 3 and 6 in Item B). Suppose that, in one case, the size of the contributions from the varieties (equivalently, the variance of the variety contributions) is larger than for those from the locations, variety-location combinations, or noise, but, in another case, it is not. Surely, a researcher would choose the first case for follow-up research on genetic factors (Neisser et al. 1996; see also Nuffield Council on Bioethics 2002, chap. 11). Similarly, if the location part of the variance were smaller than the residual or noise part, might a researcher not follow up by investigating the uneven experiences of siblings within families?

There are three problems with this line of thinking. The first two have already been noted: the values of contributions are conditional on the specific set of varieties and locations and, in human studies, the standard assumptions do not allow proper estimates of contributions associated with varieties, locations, or variety-location combinations. The third problem is that the apparent separation of variety, location, and variety-location contributions is an artifact of using the additive model as an approximation for unknown dynamics (Item F). Because the approximation is not guaranteed beyond the rerun situation, analyses based on additive models do not allow us to compare the sizes of separate contributions *in general.* This means that, even if we ignored the shortcomings reviewed in this section and persisted in labeling the variety and location contributions as (nonmeasurable) "genetic" and "environmental" factors, the relative size of these factors in the circumscribed rerun situation provides no warrant for comparisons of greater generality about genetic versus environmental influences (see end of Gap 3 in Item B).

In agricultural and laboratory settings, if breeders base their expectations and decisions on the relative size of contributions (or variances of contributions), they can compensate for any shortfalls (Approach 1). Such compensation is not an option in human studies, where selective breeding is not undertaken. Nevertheless, reference to the relative size of the fractions of variation for human traits is common. For example: "[t]he impact of genetic differences appears to increase [relative to environmental effects] with age" (Neisser et al. 1996); "most studies of intelligence, personality, and behavior turn up few or no effects of the shared environment" (relative to shared environment) (Pinker 2004, p. 10); and a trait with high heritability is a good candidate for using molecular genetic techniques "to identify the specific DNA sequences responsible for genetic influence" (Plomin and Asbury 2006, p. 87; see also Nuffield Council on Bioethics 2002, chap. 11). Recall, to say no to nature-nurture is to reject the relative weighting of genetic and environmental contributions.

f) *Combine measurable factors with partitioning of variation.* If, despite the preceding shortcomings, the relative sizes of the fractions of variation are used to inform our choice of cases in which to search further for genetic and environmental factors underlying traits, the obvious next step would be to collect and analyze data on measurable genetic and environmental factors in relation to the trait. This direction of research is the focus of Item I. An intermediate step would be to combine data on variation among varieties and locations with data on measurable genetic or environmental factors. As will be evident in the next item, this step is often taken without acknowledging what has been described in this section, that is, the shortcomings of the moves that attempt to ascribe meaning to heritability studies beyond the rerun situation. As will be seen in the following item, the distinction between traits and underlying measurable factors (Gap 2 in Item B) is often blurred.

Item H. Combine measurable factors with data on variation among varieties and locations, blur the trait-underlying factor distinction, or both: Some critical reworkings

In which lessons are drawn from reworking the frameworks of proponents and critics of heritability studies who either combine measurable factors with data on variation among varieties and locations, blur the trait-underlying factor distinction, or do both.

The gene-free formulation of Items D and E highlights the distinction between traits and underlying measurable factors (Gap 2 in Item B) and reminds us that translation from descriptive quantitative genetics analyses of traits to hypotheses about causal factors is far from direct (Gaps 3, 6, and 7). The trait-underlying factor distinction is, however, commonly forgotten or blurred, even by scholars who have emphasized that very distinction. Important implications for the interpretation of heritability studies, primarily of human traits, can be drawn through reworking the frameworks of proponents and critics who combine measurable factors with data on variation among varieties and locations, blur the trait-underlying factor distinction, or both.

1. "The Analysis of Variance and the Analysis of Causes" revisited

The distinction between statistical patterns in the analysis of a trait and measurable underlying factors seems less significant if a gradient of a measurable genetic factor (or composite of factors) is assumed to run through the differences among variety means (Gap 2, c. in Item B; Figure B.2). Such an assumption discounts the difficulties of moving from the observation of traits that differ across varieties and locations to hypotheses about measurable genetic and environmental factors that correspond to those differences, and then to the investigation of those hypotheses (Gaps 3, 6, and 7). To affirm the trait-underlying factor distinction, this section revisits some oft-cited thought-experiments and examples from agriculture and population genetics that

Lewontin (1974a, 1982) uses to critique the use of heritability and Analysis of Variance (ANOVA) in human quantitative genetics. Some errors in the original account arise from *not* keeping contributions in ANOVA (i.e., effects) and measurable factors clearly distinct.

Lewontin (1982, 132–133) introduces two agricultural thought-experiments to help nonspecialist readers visualize why heritability within groups is not relevant to an explanation of differences between groups (Gap 8). In one example, two sets of seeds sampled from one open-pollinated variety are planted in two pots of washed sand. Both pots are fed with plant-growth solution, but the solution in the second pot lacks nitrogen. Lewontin observes that, because each location (pot) is uniform, variation within each location will be associated with genetic differences among the sampled seeds. Lewontin calls this a heritability of 1. The correct value of *within*-location heritability is 0 because a new sample of seeds grown in the same location would have no correlation with the first sample. Although all the within-location variance is associated with genetic variability—or, more strictly, with differences in the ways that genetic factors in the different seeds interact with the same environmental factors in the pots—this variability is noise or residual variance in the ANOVA based on equations 1 and 2 in Item D (Table H.1). (If we thought of the sample of seeds as a set of *different* varieties and if we could clone each seed and replant it, then, because there are no replications of these varieties in each pot, noise variance is zero and the heritability within locations would then be 1.)

Table H.1 Numerical analysis of Lewontin's example of one variety, two pots*

			Estimates of contributions		Variance & heritability estimates	
Location	1	2	m	3.15	L	1.95(75%)
Variety			l_1	1.35	E	0.66(25%)
1	5, 4, 6, 3	2,1.8, 2, 1.4	l_2	-1.35		0.7225 (29%)
			e_{jk}	varied	h^2 within locations	0
					h^2 across both locations	0

* Arbitrary units estimated from the published figure.

Similarly, the heritability *across* both locations is 0. This value seems consistent with Lewontin's observation that the difference between locations— meaning between the mean measurements for locations—is "totally environmental" (i.e., entirely associated with the environmental difference of nitrogen versus no nitrogen). However, the location fraction of the variance is only .75. The only way to increase this value—that is, for the difference between the mean measurements for each location to be more strongly associated with the location-to-location (pot-to-pot) difference—is to reduce the within-location or residual variance that is associated with variability among seeds in the open-pollinated variety (see above) and with the fine-scale variation in soil conditions experienced by each replicate.

Precision in addressing these technical issues helps remind us that heritability is derived from observations made without control over underlying genetic or environmental factors (see Puzzle 3 and Gap 2). When Lewontin uses his example to make conceptual points, he accepts without comment the *control* that his example presumes over which varieties to select and grow and the ability to replicate environmental factors. Such control characterizes experimental crop or laboratory trials, but is much less so in observational crop trials and is not available in human behavioral genetics. The example also discounts the distinction between observational and experimental crop trials; it is as if human behavioral genetics could use experiments to generate knowledge of measurable factors.

In Lewontin's other example (1982, 132), one seed from each of two inbred varieties is planted in a series of pots of soils taken from different locations (Figure H.1). Lewontin observes that for each of the inbred varieties there are no genetic differences across the locations (pots) and all the location-to-location (pot-to-pot) variation must be *environmental*, that is, corresponding to differences in the soils. At the same time, noting that variety 1 does better than (or as well as) variety 2 in each location, Lewontin asserts that this gap is entirely *genetic* because the varieties experienced identical sets of locations.

The meanings of the terms *genetic* and *environmental* are ambiguous in this context. Again, it is instructive to be precise. First, notice that the advantage of variety 1 over variety 2 varies from one location to the next. In terms of ANOVA, there is a difference between variety contributions (i.e., between the means across all locations of variety 1 and 2), and there are variety -by- location-

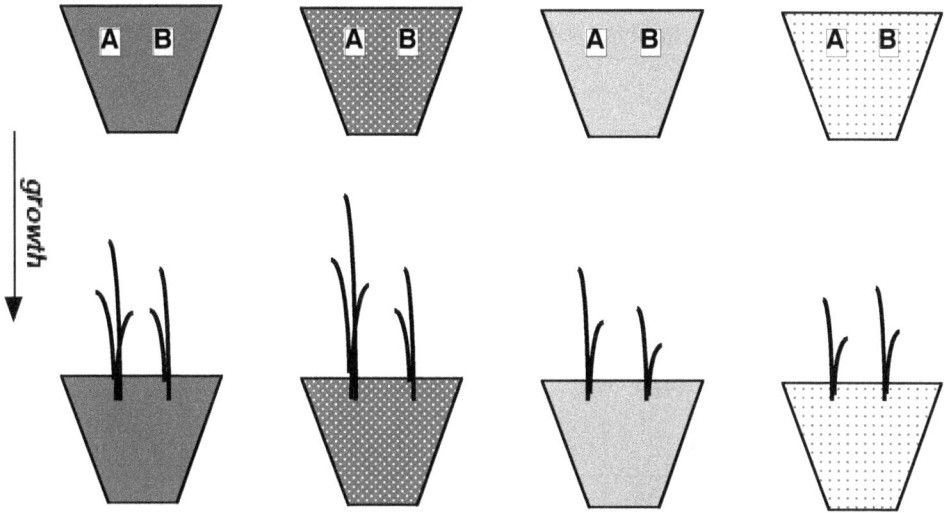

Figure H.1. Variation in two inbred varieties, A and B, grown in a series of pots of soils taken from different locations (adapted from Lewontin 1982, 132).

interaction contributions. In terms of measurable genetic and environmental factors, no hypotheses are obvious. What we can infer, however, is that the different within-location, between-variety differences are associated *either* with a different mix of genetic factors in different locations *or* with the same genetic factors having a different influence. The within-variety differences across locations may correspond with one or many environmental factors, and these factors need not be the same from one variety to the next. In short, the best we can say about the gap between varieties in each location and the gap between locations for each variety is that they are associated with some combination of unknown environmental and genetic factors.

The absence of replication means that *within-variety* heritability is not a very interesting quantity in this case. Within one location, this heritability must be 1; across locations it must be 0. (This would be the case even if the varieties were not inbred.) On the other hand, taking both varieties into account, the heritability across locations is greater than zero and L/Y is less than 1 (Table H.2)—not, respectively, 0 and 1 as might naively be expected if these heritability and among location measures corresponded to the terms *genetic* and *environmental* as used by Lewontin and others following him. In summary, this example, like the preceding one, shows how distinctions get blurred in ways that lend plausibility to lines of thinking in which measurable genetic and environ-

Table H.2 Numerical analysis of Lewontin's example of two varieties, four pots*

					Estimates of contributions		Variance & heritability estimates	
Location	1	2	3	4	m	2.75	L	2.56 (70%)
Variety					v_1	.75	V	0.56
1	4	7	2	1	v_2	-.75	VL	0.56
2	3	3	1	1	l_j	.75, 2.25, -1.25, -1.75	$h^2_{across\ locations}$	.15
					vl_{1j}	-.25, 1.25, -.25, -.75		
					vl_{2j}	$-vl_{1j}$		

* Arbitrary units estimated from the original published figure.

mental factors are separable and in which insight about those factors could follow from learning the size of estimates of heritability and the between-location component of variation.

In an earlier, much-cited essay, "The analysis of variance and the analysis of causes," Lewontin (1974a) argues in effect that, because any ANOVA is conditional (see Gap 2, d. and Item F), it cannot shed light on causes beyond the local combination of genetic types and locations observed. He supports this argument with diagrams of *norms of reaction* that summarize the response of a variety or genetic type when some environmental factor is varied. Norms of reaction for different varieties that vary in slope and position can confound any attempt to extrapolate the relative ranking of genetic types (or varieties) observed over part of the range of the environmental factor to what is observed over the full range.

Most of Lewontin's diagrams are schematic, with the measured trait plotted against an unspecific environmental factor E, but one diagram shows real data on viability of strains of fruit flies plotted against temperature. Having a single continuous environmental factor as the horizontal axis in both the schematic and real cases reinforces (*contra* Gap 2) the idea that location effects in an ANOVA can be readily translated into environmental factors with continuous gradients. Population geneticists who study fruit flies have, like agricultural breeders but even more so, control over genetic types and environmental conditions and can readily envision generating such plots. However, using diagrams of norms of reaction to make conceptual points about the ANOVA in relation to heritability studies steers us away from visualizing the difficulties in using ANOVA to expose equivalent environmental factors in humans.

2. Genotype-environment correlation: Technical versus colloquial meanings

"*Genotype-environment correlation*" (or "gene-environment correlation") has a well-defined technical meaning in quantitative genetics (Jacquard 1983; Lynch and Walsh 1998, 47). It is also used in more colloquial discussions that assume that exposure to environmental factors depends on an individual's genotype (Plomin et al. 1977, Sesardic 2005, Wikipedia n.d., a). In discussing and contrasting the two meanings, it is helpful to keep the trait-underlying factor distinction clear (Gap 2).

On the *technical* side: Genotype-environment correlation is, in the terminology of this book, variety-location correlation (or covariance). This is distinct from the variety-location-interaction variance of equations 1 and 2 under Item D. The meaning is best explained by referring to the general case of an agricultural evaluation trial, in which the means of varieties, locations, and variety-location combinations can be estimated. For the full data set, the usual method of estimation of contributions in an additive model, such as equation 1, ensures that the variety and location means (averaged across, respectively, all locations and all varieties) are uncorrelated. Within a *subset* of the full data, however, those same means can be positively or negatively correlated; this is the variety-location correlation. The definition and estimation of this quantity makes no reference to the genetic and environmental factors underlying the trait.

In human studies, varieties are typically raised in one or two locations (e.g., identical twins raised together or apart). Even in adoption studies where a number of siblings are raised in separate families, the variety and location means across *all* locations and varieties are unknown. Variety-location correlation is thus difficult to estimate (Jacquard 1983), and, if estimated (e.g., Otto et al. 1995), requires many additional assumptions (Lynch and Walsh 1998, 142).

On the *colloquial* side: It is easy to identify processes for humans through which people respond to observed *traits* of a child or through which children's *traits* lead them to seek out certain environmental factors. For example, as mentioned earlier (Item C.1), children who show an early interest in reading may be more likely to be given books and receive encouragement for their reading and book learning. The challenge for researchers has two parts: a) collect data to demonstrate that relationship; and b) develop methods of data analysis to discriminate among competing models of the association between traits and environmental factors (e.g., reactive versus active; Plomin et al. 1977, Wikipedia n.d., b).

On the other hand, the idea that exposure to environmental factors depends on an individual's *genotype* is more difficult to demonstrate. To begin with, the idea is ambiguous. Genotype-environment correlation (in the colloquial not technical sense) could have one of several meanings: a) that the exposure varies among varieties; b) that the exposure depends on traits that show substantial heritability; c) that it depends on the genetic factors possessed by the variety; or d) that it depends on the genetic factors underlying the trait. This ambiguity is evident in statements about human development of the kind "other people react to children with genotypically higher IQ by… imposing on them more intellectually demanding conversation and otherwise challenging their ability even further" (Sesardic 2005, 93). Suppose that we consider this statement using meaning d. by construing "genotypic IQ" to mean that some (yet-to-be identified) genetic factors underlie the development of whatever kind of IQ test score has been measured. What could be the mechanism through which people would match environmental factors to the child's *genetic* factors, given that these are not something that people can observe? If a matching mechanism of a kind that does not involve traits that people can observe is postulated, then evidence is needed to support that postulate. Moreover, some method of data analysis is needed to be able to discriminate among competing mechanisms, e.g., between a so-called reactive or active association between the genetic and environmental factors underlying the trait in question (Plomin et al. 1977). In short, ideas and speculation about what genes can do or about what genes can make organisms (including humans) do require *assessment of those ideas by some reliable method of data analysis.* They should be put aside unless (or until) such an assessment is possible.

Notice also, that in examining the claim about gene-environmental correlation under heading d), the focus has shifted to measurable genetic and environmental factors and away from from heritability studies. Similarly, meaning b.—exposure that varies according to traits—can, whether or not the traits show high heritability, be thought of as an invitation to determine the identity of the relevant environmental factors (Item I.2). Reference to unknown underlying genetic factors adds nothing, therefore, to the analysis. If we shift back to the realm of heritability studies, the partitioning of variance presumes exposure that varies among varieties, whether or not there is variety-location ("genotype-environment") correlation in the technical sense. In other words, meanings a. and c. are also unable to add to the analysis of variety-location correlation.

3. Residual variance is not a within-family environmental contribution ("non-shared environmental effects")

When quantitative genetic analysis divides the variation into parts, a residual fraction remains after systematic variation (among variety averages, among location averages, and among variety-location-combination averages) is taken into account (see equations 1, 2, 8, and 9 in Item D). This residual fraction, E/Y, corresponds to the variance of the noise or unsystematic contributions to the trait, that is, the variance of trait differences among replicates within variety-location combinations that are unrelated to trait variation within other combinations. In contrast, human quantitative genetics over the last two decades has given E/Y the label of *nonshared environmental effect*. The common interpretation of large E/Y value is that within-family or non-shared environmental differences are large relative to the effects due to the members of a family growing up in the same location, which are labeled "shared environmental differences" (Plomin 1999).

This label has stimulated research into environmental factors that differentially affect members within the same family. Turkheimer and Waldron (2000) review this research and Turkheimer (2000) concludes that the search has not been very fruitful. Plomin (2011) himself points to modest findings yet attests to the considerable currency given to the idea that differences within families are large relative to the effects due to the members of a family growing up in the same location. (In discussions of this idea, it should be noted, authors are not always clear when *effects* is being used to refer to fractions of variance or to some causal factor.)

Invoking a large E/Y value to motivate a search for environmental factors means either a) the trait-underlying factor distinction is not being made (Gap 2), or b) the research follows the unreliable intuition that the partitioning of variation provides insight into the relative strength of the different kinds of factors underlying the development of the trait (see Gap 3 and Item G.7, e.). The label itself needs rethinking. Residual variance is a *nonshared* contribution in the sense of *not* being associated with variation between location averages, variety averages, or additional contributions from averages for variety-location combinations. Labeling it an *environmental* contribution fails to point to its two sources: a) measurement error (after subtracting any systematic differences in measurement error across varieties or across locations), which can be reduced by greater accuracy in measurement; and b) differences among replicates within variety-location combinations in the ways that the (unknown) genetic and environmental factors possessed or experienced by the replicates influence the trait. Such differences provide no basis for expecting the same kind of *environmental* factor—or the same combination of genetic and environmental factors—to generalize across locations (or families). This second source of residual variation can be reduced if replicates of a variety are more uniform and positioned randomly within the location. Random positioning might reflect something about environmental factors, but uniformity of replicates suggests something about the variety. In sum, *nonshared environmental effects* is a misleading term for the E/Y fraction of the variation; the neutral terms *noise*, *unsystematic*, or *residual* are more appropriate.

4. The "laws of behavior genetics" revisited

Turkheimer (2000) presents three laws of behavior(al) genetics (i.e., quantitative genetics for human behavioral traits). He notes confidently "that the empirical facts are in and no longer a matter of serious controversy. [I]t is time to turn attention…to the implications of the genetics of behavior for an understanding of complex human behavior and its development." This section takes a different view of the laws, the empirical claims, and their implications in light of the trait-underlying factor distinction (Gap 2), the unreliability of estimates of human heritability and other fractions of variance (Item E), and the reinterpretation of residual variance (section H.3 above). Table H.3 presents the laws and empirical facts, re-expressed as claims that keep the trait-underlying factor distinction clear and adjusted in terms of the gene-free formulation.

Table H.3. The three laws of behavioral genetics

A. Three laws of behavioral genetics
(Turkheimer 2000 [with additions in brackets from Turkheimer 2004])
First Law. All human behavioral traits are heritable.
Second Law. The [environmental] effect of being raised in the same family is smaller than the effect of genes [and is often close to zero].
Third Law. A substantial portion of the variation in complex human behavioral traits is not accounted for by the effects of genes or families.

B. Laws from A re-expressed to keep the trait-underlying factor distinction clear
For all human behavioral traits,
1. Heritability of the trait is substantially above 0.
2. Fraction of variance associated with differences among location averages is smaller than the fraction associated with differences among variety averages, and is often close to zero.
3. A substantial portion of the variation is not accounted for by the preceding two fractions of variance.

C. Laws from B restated as empirical claims
(adjusted in terms of the gene-free theory of Items D and E*, using $\mathcal{V}$ to stand for (1-2γ)V⁺/Y)
For all human behavioral traits,
1. $V^+/Y + \mathcal{V}$ is substantially above 0.
2. $L/Y < V^+/Y + 2\mathcal{V}$. Often L/Y is as low as $\mathcal{V}$.
3. E/Y is substantially above 0.

D. Empirical claims from C re-expressed to keep the trait-underlying factor distinction clear
For all human behavioral traits,
1. Broad-sense heritability is above 0, but to a degree that depends on γ and VL.
2. The fraction of variance associated with differences among location averages may or may not be smaller than the fraction associated with differences among variety averages and may or may not be close to 0; this depends on γ and VL.

Table H.3. (continued)

> 3. A substantial portion of the variation is not accounted for by differences among location averages, differences among variety averages, or differences among variety-location combination averages.

* Equations 15 and 16 in Item D indicate that the conventional estimates of "heritability" and "L/Y" given by equations 18 and 19 in Item E are actually estimates of $V^+/Y + V$ and $L/Y - V$, which expand out to, respectively, $2(1-\gamma)(V + VL)/Y$ and $L/Y - (1-2\gamma)(V + VL)/Y$. See Appendix I for numerical illustration.

The empirical claims, carefully construed, tell us that more data are needed to estimate γ and VL before reliable patterns in the V, L, and VL fractions of variation of human behavioral traits are clear. Even then, those patterns concern variation in the trait, *not the relative influence of the genetic and environmental factors underlying the development of those traits* (see Gaps 2, 3, 6, and 7). The residual variance, E/Y, as explained in section 3 above, has two sources: measurement error, and differences among replicates within variety-location combinations in the ways that the (unknown) genetic and environmental factors possessed or experienced by the replicates influence the trait. It *might* be that interesting findings about within-family environmental factors emerge from research on differences among siblings raised together (notwithstanding the assessments of Turkheimer and Waldron 2000 and Plomin 2011 mentioned earlier). However, heritability studies, when translated so that the trait-underlying factor distinction is clear and adjusted in light of the gene-free theory of Items D and E, *do not provide a justification for that line of inquiry.* Similarly, although it *might* be time to take up the challenge of examining the ways that "[b]ehavior emerges out of complex, nonlinear developmental processes" (Turkheimer 2000, 161), the warrant for that shift of attention is not provided by the empirical facts of behavioral genetics.

5. Studies of offspring of identical twins

Since Turkheimer published the work reviewed in the previous section, he, his colleague Emery, and their students, have analyzed the similarity of offspring of monozygotic twins to clarify the relationship between parental traits, especially divorce, and the behavior of their offspring. Turkheimer (2008, 4) describes the logic of their analyses in two scenarios:

[I]f a genetic propensity to be aggressive makes parents more likely to get divorced, and those same genes when passed to the children make them more likely to be aggressive on the playground, then one will observe an association between divorce and playground aggressiveness that will not really be a causal consequence of divorce....But in identical twin parents...none of the differences between the children can arise from differences in the genes of their twin parent, so if the children do differ, we can (almost...) rule out a genetic explanation of the association.

Conversely,

Suppose poor families are more likely to be divorced than well-off families, and children raised in poor families are more likely to be delinquent. [We could] observe an association between divorce and delinquency that doesn't have any causal relationship to divorce. But twin parents share their family history of poverty, so if the children of the divorced twin are more likely to be delinquent than the children of the nondivorced twin, the parental poverty isn't a plausible alternative explanation...

Turkheimer follows up the parenthetical "almost" with caveats concerning, for example, the contribution of the other nontwin parent. Not included among his caveats, however, are various points made in or implied by previous items:

a) shortcomings in estimation of human heritability (Gap 5 and Item E), presuming that heritability is what "genetic propensity" refers to;

b) unreliability of the heuristic that all other things being equal, similarity in traits for relatives is proportional to the fraction shared by the relatives of all the genes that vary in the population (Item E.2);

c) heritability ("genetic propensity") as a measure of similarity in the trait does not translate in any direct way to hypotheses that invoke underlying genes or genetic factors (Gaps 3, 6, and 7 and Items F and G); and

d) asymmetry in conceptualization of genetic and environmental (or social) factors—the latter are measurable; the former unknown and potentially heterogeneous (sections 1 and 2 in this Item).

Turkheimer (2008, 5) reports "a rich variability of outcomes"—some indicating a genetic association, some an association with the social factor, and some ruling out hypothesized associations. We do not know, however, how much this variability of outcomes is generated by the unreliable heuristic (issue b. above). Even then suppose that, for the purposes of discussion, we had results accompanied by an analysis of sensitivity to variation away from the unreliable heuristic value, or even that the heuristic were dispensed with because we had data on the necessary classes of relatives (Items D and E). Could we then interpret similarity or dissimilarity of offspring of monozygotic twins as a "reflection of the environmental and genetic developmental processes that underlie complex human behavior" (Turkheimer 2008, 5)? My answer: Not readily. Turkheimer's phrases "genetic propensity" and "genetic explanation" suggest an equivalence or a direct translation between measures of similarity in the trait and hypotheses that invoke genetic factors underlying the trait, but this is not so (issue c. above and Gaps 2, 3, 6, and 7 in Item B). "Genetic explanation," moreover, suggests a symmetry with environmental or social explanation, and this is not so either.

The asymmetry can be seen by examining the first scenario used by Turkheimer to describe the logic of the analyses followed by an alternative scenario, and by considering what interventions to modify the developmental processes follow from the analyses. In the first scenario there are no grounds for a "genetic explanation" of the association between divorce and the outcome—aggression of child. Table H.4 summarizes this scenario in its simplest form. (Notice that the nonmeasured, unknown genetic factor or composite of factors contributing to the similarity of the *first* pair of twins for the trait—here, aggression—is shown as distinct from the genetic factor contributing to the similarity of the *second* pair of twins, given that the analysis cannot show them to be the same or similar.) To reduce the incidence of aggression seen in children of divorced parents, it would be plausible under this scenario to focus on policies to reduce divorce without reference to genetic factors or genetic relatedness.

Now imagine an alternative scenario, shown in Table H.5, in which the analysis does *not* rule out a "genetic explanation" of the association between divorce and the outcome, aggression of child.

What is learned from the alternative scenario about possible interventions to modify the "environmental and genetic developmental processes that underlie complex human behavior"? This scenario would weigh against the social conservatives' focus on policies to directly reduce divorce, such as covenant marriages or at-fault divorce. But beyond that, what? If we sought to identify the unknown genetic factors associated with aggression of a parent, nothing in the analysis rules out these factors differing from twin pair to twin pair (Puzzle 2 in Item A) and among non-twinned parents. In light of that possibility, we

Table H.4. Analysis of offspring of monozygotic twins: Behavioral outcome rules out a "genetic explanation" of the association between divorce and the outcome, aggression of child.

Offspring of	Behavioral outcome (aggression of child)	Factors influencing behavioral outcome			
		Nonmeasured, unknown genetic factor (or composite of factors) associated, as measured by heritability, with aggression of twin parent		Environmental factors	
				Aggression of twin parent (related to hypothetical genetic factors)	Divorce of parents
		g_1	g_2	e_1	e_2
Twin1 of 1st pair	Yes	[hatched down]		[hatched up]	[black]
Twin2 of 1st pair	No	[hatched down]		[hatched up]	[white]
Twin1 of 2nd pair	Yes		[hatched down]	[hatched up]	[black]
Twin2 of 2nd pair	No		[hatched down]	[hatched up]	[white]

Notes: Black/white cells indicate presence/absence of the factor. Diagonal hatching downwards denotes offspring of any identical twin are not identical in their genetic factors to that twin parent because there is a second parent. Similarly, offspring of the different identical twins are not identical in their genetic factors. Diagonal hatching upwards denotes twin parents are not identical in the measured factor because heritability is less than 1.

might restrict our focus to close relatives (Gaps 3 and 6, path c. in Item B). In the case of identical twins who become parents, once aggression is seen in the children of one twin, we could advise the other twin to be more attentive to the issue of aggression. That is, we could seek ways to help the parent reduce their own aggression insofar as it affects the children (starting perhaps before the parents have offspring). And we could seek ways to help the children reduce their aggression. However, we could also advise such attention to aggression *independently of knowledge about genetic relatedness and of hypotheses about underlying genetic factors.* It would only be helpful to establish that (unknown)

Table H.5. Analysis of offspring of monozygotic twins: Behavioral outcome does not rule out a "genetic explanation" of the association between divorce and the outcome, aggression of child.

Offspring of	Behavioral outcome (aggression of child)	Factors influencing behavioral outcome							
		Nonmeasured, unknown genetic factors (associated, as measured by heritability, with aggression of twin parent)						Environmental factors	
								Aggression of twin parent (related to hypothetical genetic factors)	Divorce of parents
		g_1	g_2	g_3	g_4	g_5	g_6	e_1	e_2
Twin1 of 1st pair	Yes	▨						▨	■
Twin2 of 1st pair	Yes	▨						▨	
Twin1 of 2nd pair	No		▨						■
Twin2 of 2nd pair	No		▨						
randomly chosen parent A	Yes			▨				▨	
parent B	No				▨				
parent C	Yes					▨		▨	■
parent D	No						▨		■

Notes: In any study, offspring will come from several twins and randomly chosen parents, but the number shown is sufficient to illustrate the logic of the scenario. Black/white cells indicate presence/absence of the factor. Diagonal hatching downwards denotes offspring are not identical in their genetic factors to the twin or randomly chosen parent because there is a second parent. Similarly, offspring of parents that include one or the other of an identical twin pair are not identical in their genetic factors. Diagonal hatching upwards denotes a) twin parents are not identical in the measured factor because heritability is less than 1; and b) similarly, for non-twin parents, the hypothetical genetic factor is not expressed fully as the measured environmental factor.

genetic factors had an influence if that also meant that environmental interventions in the developmental processes would be unlikely to succeed or would be a diversion of resources from measures more likely to be fruitful. There is no justification for assuming that is so.

In sum, in an analysis of the similarity of offspring of monozygotic twins, the environmental factors are measurable and point to interventions, but the genetic factors are unknown, potentially heterogeneous, and informative only for advising close relatives, even in the thought-experiment where issues a. and b. have been overcome. (Of course, behavioral outcomes and environmental factors may also be heterogeneous [Gatze-Kopp et al. 2012], but this does not help us interpret the outcomes of the analyses of Turkheimer and colleagues.) This asymmetry, it should be noted, can be found in a larger set of studies that attempt to inject heredity (Kendler and Baker 2007) into the longstanding "social causation" versus "social selection" debate in the sociology of mental illness (Hudson 2005). (Social causation means low socioeconomic status [SES] increases risk of mental illness, while social selection means that the mentally ill decline in SES because of their illness.) Notwithstanding the methodological difficulties in distinguishing the two kinds of pathways in a given data set, heritability (link 1 in Figure H.2) again points to unknown, possibly heterogeneous genetic factors while the environmental factors are measurable and open to intervention (links 2, 3, and 4).

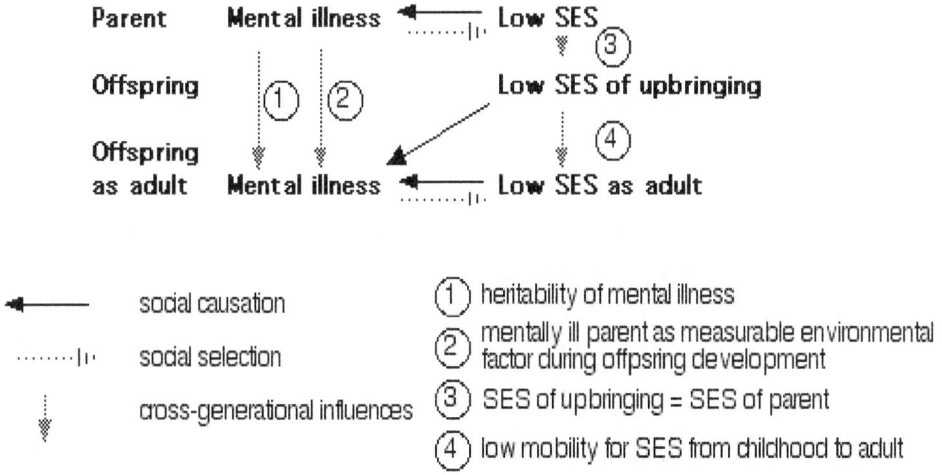

Figure H.2 Pathways of social causation and social selection for socioeconomic status (SES) and mental illness.

6. "Missing heritability" and the possible heterogeneity in factors underlying the development of a trait

As mentioned earlier (Item C.2), genomic studies have had difficulty identifying causally relevant genetic variants behind variation in human traits (McCarthy et al. 2008, Couzin-Frankel 2010). Even when many genetic variants are examined together, only a small fraction of the variation in the trait is associated with—or in statistical terms, "accounted for" by—the genetic variants. This finding has led to discussions about *missing heritability* (e.g., Manolio et al. 2009). The expectation that genetic variants could account for a larger fraction of trait variation follows from ambiguous descriptions of heritability as, to use a previously cited example, the "fraction of the variance of a phenotypic trait in a given population caused by (or attributable to) genetic differences" (Layzer 1974, 1259). To remind ourselves of the trait-underlying factor distinction, "new heritability" could be defined as the fraction of variation associated with the summation term in following equation:

$$y_i = \Sigma\, b_s\, gf_{is} + e_i \tag{27}$$

where y_i denotes the measured trait y for the ith individual;

gf_{is} is the value of genetic factor s for the ith individual;

e_i is an unsystematic or noise contribution adding to the trait measurement;

b_s is a coefficient for genetic factor s estimated so as to minimize the variance of e_i's.

There is no obvious mapping of equation 27 to equation 1 (Item D), which adds up contributions from varieties, locations, variety-locations combinations, and noise, and thus allows heritability, V/Y, to be estimated.

We might expect a higher fraction of variation to be accounted for if environmental factors were also included:

$$y_i = \Sigma\, b_s\, gf_{is} + \Sigma\, b_t\, ef_{it} + \Sigma\Sigma\, b_{st}\, gf_{is}ef_{it} + e_i \tag{28}$$

where ef_{it} is the value of environmental factor t for the ith individual;

b_s, b_t, b_{st} are coefficients estimated so as to minimize the variance of e_i's.

There is still no obvious mapping of equation 28 to equation 1. In short, new heritability has no empirical or conceptual relationship to the heritability of heritability studies. (Item I.4 provides further discussion of the fraction of variation associated with genetic variants in the context of the possibility of heterogeneity in the genetic and environmental factors underlying a trait.)

*   *   *   *   *

The introduction to Part II stated that discussion would be minimized of the kind "X says this, but under *my* account that should be corrected to say…" Item H has, however, included author-specific critique. Ironically, that follows from the very reason I gave for avoiding it, namely, I would want first to know whether and how other authors have addressed each of the gaps. If they had acknowledged the trait-underlying factor distinction (Gap 2) at all times, I might have had little to criticize, because acknowledging that distinction would have required them to be clear about the different meanings they were giving to key terms, such as *genetic* and *environmental* (Gap 1). Indeed, my sense is that heritability studies would engender less debate all round if researchers admitted that they really want to be measuring genetic and perhaps environmental factors underlying the trait in question or even discussing the pathways of development that such measurable factors modulate (Turkheimer 2004, Keller 2010, and Tabery 2009, 2014). With such an admission, researchers would not need to use ambiguous terms (Gap 1), blur the trait-underlying factor distinction (Gap 2), discount the difficulties in translation (Gaps 3, 6, and 7) and in extrapolation (Gap 4). They might be comfortable considering alternatives to the standard assumptions in human heritability studies (Gap 5 and Item E) or even leaving behind any interest in the relative weighting of nature and nurture.

Explicit reference to the genetic and environmental factors underlying a trait is taken up in Part III. To close out Part II, let us take stock of where the last two items have brought us. By reworking the frameworks of proponents and critics who combine measurable factors with data on variation among varieties and locations, blur the trait-underlying factor distinction, or do both, these items reinforce the conclusion of the Short Critique: There is almost nothing reliable that we can do with the information about the similarity among relatives that is the basis of heritability studies (a field that fits into Approach 5 of Item G) *unless we can revert back to one of Approaches 1–4*. These approaches can be made to work in situations that are possible in agricultural and laboratory research, situations that (with a few exceptions noted in Item G.3 and G.4) do not apply to studies of humans.

PART III

POSSIBILITIES, PROBLEMS:
SPACES OPENED UP
FOR FRESH INQUIRY

Introduction to Part III

Parts I and II have shown that there is almost nothing reliable that we can do with the information about the similarity among relatives that is the basis of heritability studies—unless we have control over materials and situations in ways that are possible in agricultural and laboratory research. In studies of humans, such control, for the most part, does not apply, so in that domain it makes sense to leave heritability studies behind.

Human quantitative geneticists will surely resist this conclusion. Possible objections are summarized, with my counter-responses, in Appendix 4. Part III, however, leaves such exchanges behind. After all, saying *no* to nature-nurture by no means leaves us empty handed. We can turn to research on the specific, measurable genetic and environmental factors underlying human traits in which no reference is made to a trait's heritability or the other fractions of the total variance. There is a wealth of such research (albeit less on unraveling the influences on both sides together than on investigating one or the other side, as in the research programs of *genetic* epidemiology and *social* epidemiology). Part III does not attempt a comprehensive review of research associating human traits with measurable factors, but considers various possibilities and problems of different research programs outside heritability studies. To this end it takes up the themes developed in Part II concerning heterogeneity and causality grounded in practice. The message is that moving beyond *no* on nature-nurture leads us to *maybe* for genetic and environmental.

The two items that make up Part III are offered in the spirit of unfinished and open inquiry, aiming to encourage readers to make further contributions of their own to discussion of heredity and variation. Item I considers five kinds of studies of measurable genetic and environmental factors underlying human traits, making use of the framework introduced in Item G to examine what is possible once the possibility of heterogeneity in genetic and environmental factors is taken into account. In particular, we can consider what actions are entailed in mimicking or exceeding the control of variety and location that are often feasible in agricultural research. Item J extends beyond a focus on heredity and variation to sketch four kinds of study that might allow for the heterogeneity of any kind of factors underlying patterns in observed traits or variables in the human and social sciences.

Let me note one line of inquiry not taken up. To the extent that newer research on genetic and environmental factors has been motivated or informed by incorrect or unreliable estimates of heritability and of other fractions of variance (Gap 5, Item E, and Item H.3), we might reexamine the methods and results of the newer research. More generally, do similar or analogous oversights to those around classical heritability studies (discussed in Parts I and II) limit the level of progress that is possible with the newer methods? Such a reexamination lies, for the most part, beyond the scope of Part III; my mentioning it here is meant as another invitation to interested readers to expand on what this book provides.

Item I. Heterogeneity, control, and research that focuses on measurable genetic and environmental factors underlying human traits

In which lines of research on measurable genetic and environmental factors underlying human traits are reviewed, with special attention paid to the possibility of underlying heterogeneity and to the control of conditions needed for the analysis of the data or for the application of the results.

This item reviews research on the variation and inheritance of human traits that omits reference to the key quantities of heritability studies, namely, the trait's heritability and the other fractions of the total variance. Five lines of research are discussed in relation to two considerations: a) the possibility of underlying heterogeneity; and b) the control of conditions that is implicated in the analysis of the data or the application of the results.

First, let us generalize the possibility of underlying heterogeneity (see Figure A.2) beyond heritability studies: When similar responses of different *individual types* are observed, it is not necessarily the case that similar conjunctions of *measurable factors* have been involved in producing those responses. In this book we have been considering primarily genetic and environmental factors, but the issue arises across the human sciences. In epidemiology, for example, we would be considering heterogeneity in risk and protective factors. Indeed, the simplest form of heterogeneity—variation *around* the average for a group—arises whenever methods of data analysis, such as the Analysis of Variance, test hypotheses about differences *among* the averages of two or more groups.

The control of conditions—or situations—will be considered especially in relation to heterogeneity. This focus follows the lead of Item G, which examined the actions possible or proposed on the basis of what is known if one takes into account the possibility of heterogeneity in the genetic and environmental factors. The control of conditions has also been touched on at several other points (e.g., Puzzle 3 in Item A and as revisited in Item C, Gaps 4, 6, and 7 in Item B, Item H.1), but its significance will come into sharper focus through discussion of the lines of research in this item and the next.

1. PKU: Heterogeneous factors complicate the poster-child case of a genetic condition

The first line of research is not new, but is included so as to illustrate how heterogeneity and the control of conditions are related. The case of phenylketonuria (PKU), discussed in Item E.1, looks at first sight to be straightforward. A single measurable genetic factor can be detected at birth by a biochemical test. Severe cognitive impairment can be averted if the appropriate environmental factor—an adhered-to special diet with reduced levels of the amino acid phenylalanine—is present. Indeed, this possibility of treatment makes PKU the poster child for genetic medicine.

However, as Paul's history of PKU screening describes, the certainty of severe cognitive impairment has been replaced by a chronic disease with a new set of problems (Paul 2000, Paul and Brosco 2013, 111ff). For example, although screening of newborns became routine quite rapidly during the 1960s and 70s in the United States, in many states there remains an ongoing struggle to secure health insurance coverage for the expensive special diet. It can be hard to enlist family and peers to support individuals with PKU staying on the diet through adolescence and into adulthood. Math deficits common among individuals with PKU make diet calculations difficult. In short, individuals with PKU are subject to heterogeneous influences on their pathways of development over the life course. The apparent simplicity of the environmental factor presumes control over those heterogeneous influences—control that is often not fully present.

Likewise, the apparent simplicity of the genetic factor presumes control over the varying degrees of dietary modification needed by different individuals. Screening has improved over time so that false positives—individuals said to have PKU but needing no special diet—are rare (Sweetman 2001). Yet differences remain among individuals with PKU because hundreds of mutations have been identified in the PAH gene; any individual with PKU may have one pair of mutations out of hundreds of thousands of possible pairs. Efforts are underway to classify mutations as mild, moderate, or severe and as responsive to the drug BH4, which allows a higher-protein diet. Tailoring the dietary modification will require acceptance and application of the diagnostic classification of mutations, secondary screening to place individuals in the right category, and maintenance of the appropriately calibrated diet despite the issues noted in the previous paragraph and some new ones—access to BH4 and keeping individuals who take it from going off the special diet altogether, which is not medically recommended (Paul and Brosco 2013, 108-109). Conversely, when individuals with PKU are not differentiated, this presumes that the medical and health insurance system can withstand any pushback from individuals with PKU or their families about the lack of more personalized treatment.

To acknowledge heterogeneity is not simply to promote more refined genetics-based diagnosis. It can also mean getting involved at a variety of points that shape the PKU life course, such as support groups for individuals and families, campaigns for insurance coverage for the special diet, counseling, or paid family leave. As we consider more recent lines of research on measurable genetic and environmental factors underlying human traits, the case of PKU can serve to remind us to examine the actions possible or proposed on the basis of what is known if one takes into account the possibility of heterogeneity in the genetic and environmental factors.

2. Interaction of a single gene and a single environmental factor for behavioral traits

In 2002–3 Caspi, Moffitt, and colleagues published two articles in *Science* that examined psychological traits in relation to measured genetic and environmental factors (Caspi et al. 2002, 2003). Moffitt et al. (2005) subsequently laid out the research program, noting that, "in psychiatric genetics, ignoring nurture handicaps the field's capacity to make new discoveries about nature." Some meta-analyses have cast doubt on the generality of the Caspi and Moffitt results (Risch et al. 2009; see Morris et al. 2007), but in order to illustrate the approach and to raise pertinent issues, the discussion that follows uses the 2002 study. The issues raised apply to the overall research program of examining interaction between measured genetic and environmental factors even if these particular 2002-3 results turn out not to be widely replicated.

The 2002 *Science* article reports on antisocial behavior in adults in relation to the activity of monoamine oxidase typeA (MAOA) and childhood maltreatment; MAOA deficiency is a strong predictor of aggressive behavior only when the child has also been maltreated. The authors conclude that their results "could inform the development of future pharmacological treatments" (Caspi et al. 2002, 853). In the context of research on childhood experience in relation to adult behavior, the implication of that conclusion is that, if low

MAOA children could be identified, prophylactic drug treatment could reduce their propensity to antisocial behavior as adults. Or, to be more precise, such treatment could reduce their vulnerability, that is, the risk that childhood maltreatment would pave the way to undesired adult outcomes.

An easy rejoinder to the authors' conclusion would be that, if childhood maltreatment could be identified and stopped early, such action could reduce their vulnerability to low MAOA levels paving the way to undesired adult outcomes (Morris et al. 2007). Indeed, eliminating childhood maltreatment would seem to be unconditionally positive, while prophylactic drug treatment may have side effects, some of which may not emerge until later in life. The rejoinder is, however, too easy. Although the intended outcome—eliminating childhood maltreatment—may be seen as positive, the way to get to that outcome might not seem unconditionally positive. After all, detecting and preventing childhood maltreatment might require intrusion into many households, surveillance, monitoring, and intervention by state agencies, diversion of government budgets from other needs, and so on.

This last sentence speaks to the control of conditions that is implicated in the application of the results of an analysis of genetic and environmental factors. The MAOA-maltreatment case also brings in the other theme of heterogeneity underlying the variation. Notice that the points plotted in Figure I.1 are the averages for the respective categories of people. Within each category people show a range of antisocial behaviors. It turns out that, among children who experienced probable or severe maltreatment, the ranges overlap, that is, some of the high MAOA individuals ended up with higher antisocial behavior scores than some of the low MAOA individuals.

Adjustment of what counts as antisocial does not succeed in eliminating the potential for misclassifying which children are ones who may end up antisocial. If we counted as antisocial only those individuals whose score exceeded some value that is higher than the upper limit of the range for high MAOA individuals, this would increase the numbers of low MAOA individuals who did not end up counting as antisocial. If we lowered the cutoff score, many high MAOA individuals would end up with behavior classified as antisocial (Figure I.2). Indeed, by playing around with the cutoff score, the best that can be achieved with Caspi et al.'s 2002 data is to classify correctly a little more than one-third of children on the basis of their MAOA status (i.e., low MAOA ending up antisocial and high MAOA not ending up antisocial).

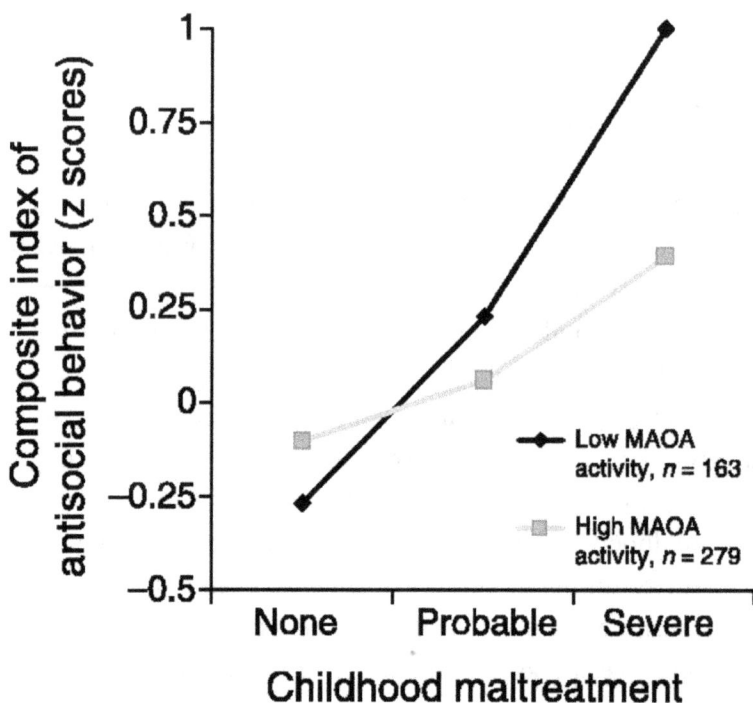

Figure I.I. Average adult composite antisocial behavior score in relation to levels of MonoAmineOxidaseA and level of childhood maltreatment for a sample from Dunedin, New Zealand (from Caspi et al. 2002, 852, reproduced with permission).

The issue of misclassification is especially troubling because, once the resources are invested to screen children for MAOA levels, attention would be focused on *all* low MAOA children. Indeed, how could treating children according to their genetic group be avoided if we do not know from a childhood MAOA assessment whether any particular individual is one who would go on, after maltreatment, to become an antisocial adult? Parents might push for additional research to identify other characteristics that differentiate among the low MAOA children and perhaps help predict who among the high MAOA children are also vulnerable. If that research took place and were successful, additional resources would have to be invested to customize the way that parents, teachers, doctors, and social workers treated the different low- and high-MAOA children and to educate everyone not simply to treat children according to which MAOA group they were a member of. In short, the implications of results based on the interaction of genetic and environmental

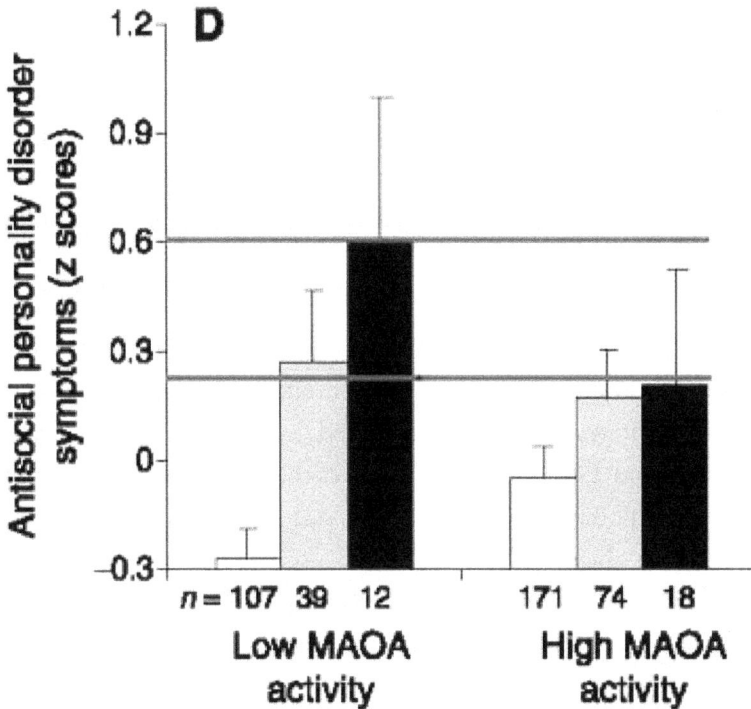

Figure 1.2. Average and standard deviation for adult antisocial personality disorder score in relation to levels of MonoAmineOxidaseA and level of childhood maltreatment for a sample from Dunedin, New Zealand (from Caspi et al. 2002, 852, reproduced with permission). Horizontal lines indicate the misclassification effect of a low and high cutoff for labeling a person as antisocial (added by author; see text for discussion). The width of the bars is not proportional to the numbers of people, n.

factors depend on what action is proposed once the possibility that heterogeneity underlies those factors is taken into account.

Now, suppose the MAOA-maltreatment data concerned not antisocial behavior but a less charged trait, say, a specific adult disease. What kinds of medical diagnoses would receive the necessary investment in pharmaceutical and sociological research, screening, and preventative treatment or monitoring to address the conjunction of genetic and environmental factors involved? One answer is that, especially with the efforts of well-organized parental advocacy groups (Panofsky 2011), government funding might be secured to address the prenatal diagnosis or post-natal treatment of rare debilitating genetic disorders such as PKU (section 1). Another answer is that the most likely focus for public and corporate policy would be on diagnoses for which the net benefit, i.e., the number of vulnerable people times the average benefit of ameliorating the effect of the genetic difference, would be large. (More precisely: the focus would be on diagnoses for which the benefit minus the cost of research, screening, and treatment is largest.)

What would the MAOA case lead us to expect for medical diagnoses with large benefit/cost ratios? If the effect of some genetic difference depended on identified social or environmental factors, and if variability within the groups that have, on average, high and low vulnerability produced a problem of misclassification, pressure might arise to differentiate among individuals within the groups. Until additional research succeeded in identifying distinguishing characteristics, we would expect that parents, teachers, doctors, social workers, insurance companies, policy makers, friends, and the individuals themselves would do the best that they could, which is to use the genetic information to *treat individuals according to their group membership*. If the additional research were not conducted or were not successful, we might never get beyond treatment according to group membership. Here is where the control of conditions implicated in the application of the results becomes relevant: Whether attention gets paid to the heterogeneity—which, in this case, takes the form of variation around means—would depend on decisions about the cost of differentiating among individuals and on the amount of pressure that members of the public created to differentiate among individuals within the groups.

This scenario is relevant to the prospect of personalized medicine (Wikipedia n.d., c), which, in its simplest form, involves the use of genetic information to predict which patients with a given condition (e.g., heart arrhythmia) will benefit from a particular drug treatment (e.g., beta blockers). More ambitiously, personalized medicine promises to inform people of their heightened vulnerability (or resistance) to specific environmental, dietary, therapeutic, and other factors early enough that they can adjust their exposure and risky behaviors accordingly. As the analogy to the MAOA case indicates, the path to *personalized medicine* will involve a phase in which large numbers of people are treated *according to their group membership*. Moreover, this phase may not be a passing one. The open question—another puzzle—is to understand the social conditions in which the information and resources needed to move beyond the group membership phase would be forthcoming.

3. Interaction of many environmental factors for behavioral traits

Kendler and colleagues examine behavioral traits in relation to a wealth of environmental factors over the life course as well as to the relatedness of the individuals (Kendler and Prescott 2006). In Kendler et al. (2002), for example, data on over 1,900 twins are used to fit the incidence of major depression to an additive model that incorporates many environmental factors and a "genetic risk" factor. This last factor is derived from the incidence of major depression in the co-twin and parents, with adjustments made for the degree of relatedness of the twins (monozygotic versus dizygotic). The model accounts for 52% of the variance in the trait and provides a picture of development that is rich and plausible (Table I.1). For example, the path coefficient of .7 from neuroticism (NE) to low self-esteem (LS) and of .3 from LS to low education (LE) suggests that neuroticism makes it more likely that a person has low self-esteem and that, in turns, makes it more likely that they do not pursue education as far as others.

Kendler et al. (2002, 1133) show admirable reserve about how to interpret the fitted model, concluding that their "results…should be treated with caution because of problems with causal inference, retrospective recall bias, and the limitations of a purely additive statistical model." Let us, however, delve further. One other limitation of a purely additive model, introduced in Item F.5, is that the values of the correlations or path coefficients are dependent on the network of factors chosen by the researchers. For Kendler and colleagues, this dependency might not seem to be a noteworthy limitation, given, as the paper's title—"towards a comprehensive developmental model"—indicates, a very wide range of factors are taken into account. Moreover, in the initial network every factor is linked to every other factor. The links are then pruned to arrive at a less saturated network, but it is the data analysis that indicates which links in the network have low correlations or path coefficients. Finally, before arriving at the final published network, the researchers examine whether the degree of fit of the model to the data is sensitive to the inclusion or omission of factors.

The preceding virtues notwithstanding, the results remain dependent on the network of factors chosen by the researchers. The research of Kendler et al. does not include factors that correspond to therapeutic interventions or to social changes that have led to the rising incidence of depression. (Kendler [pers. comm.] explains that their research is intended to be basic, not applied, science.

Table I.I. Model for Predicting an Episode of Major depression*

Upstream variable (rows) × *Downstream variable* (columns)

Upstream \ Downstream	GR	DF	CS	CP	NE	LS	EO	CD	LE	LT	LS	SM	ED	PH	MP	DI	DS	IS	EM
GR		.3	.2	.2	.1				.1			.1	.1	.1					.1
DF	.3		.3	.3	.2	.1	.2	.3	.1	.4					.1	.1			
CS	.2	.3		.2	.1		.1	.2	.1	.2		.2				.1		.2	
CP	.2	.3	.2						.1										
NE						.7	.4				.2				.2				.2
LS									.3					.1					
EO								.1		.1	.1	.2		.3				.1	.1
CD									.1	.2	.3								.1
LE										.1		.2	.2						
LT											.1	.3	.1		.1			.1	
LS												.2							
SM														.1	.2		.2	.2	
ED														.2	.1				
PH																.1			.1
MP																.1	.2		.1
DI																			.1
DS																			.4
IS																			.2

* Adapted from Kendler et al. (2002). The numbers represent the correlation or path coefficient between the variables: GR, genetic risk for major depression; DF, disturbed family environment; CS, childhood sexual abuse; CP, childhood parental loss; NE, neuroticism; LS, low self-esteem; EO, early-onset anxiety; CD, conduct disorder; LE, low education; LT, lifetime trauma; LS, low social support; SM, substance misuse; ED, ever divorced; PH, past history of major depression; MP, marital problems in last year; DI, difficulties in last year; DS, dependent stressful life events in last year; IS, independent stressful life events in last year; EM, episode of major depression in last year. See text and original source for further explanation.

They want to elucidate the etiology of depression in an individual's lifetime, not to get involved in debates about therapy or public policy.) Without assembling data on such factors, there can be no analysis of the sensitivity of the analysis to inclusion or omission of those factors from the model. Among individuals similar in the factors that are included in the model there may have been heterogeneity in the factors left unmeasured and excluded from the data analysis. In other words, it is possible to ignore the possibility that heterogeneity might underlie the included factors by discounting actions of a therapeutic or policy nature. Moreover, this is not the only way that what becomes accepted knowledge is linked to actions, or, in this case, *in*action. Even if readers interpreted the research the way Kendler intended and refrained from drawing therapeutic or policy implications, the analysis is only relevant if developments in therapy or policy subsequent to the period for which the data were collected have no effect on the etiology of depression or on risk factors implicated in that etiology.

Let me qualify this last conclusion and digress for a few paragraphs before returning to the analysis of Kendler and colleagues. As an alternative to making an assumption about developments beyond the period and beyond the population studied, the fitted model could be viewed as a snapshot of one population at one time. Indeed, another limitation of additive models is that the can be viewed as providing only rerun predictability. This concept, introduced in the context of heritability studies (Item F), can be extended beyond that context through the following steps: We assume some unsystematic noise in the unknown dynamics through which factors influence the trait. Then we imagine that the same sets of factors are observed again where the only changes in this rerun are noise at the same level as the original, but uncorrelated with it, and small differences in some or all of the factors. Provided that the noise is small and the rerun remains close to the original situation, a good approximation for a wide range of actual, but unknown, dynamics can be provided by additive models, such as the one depicted in Table I.1.

Exponents of additive models of measurable factors do not, in general, portray their analyses as providing merely rerun predictability. They try to work around the limitations of the rerun conditions by construing measurable factors in causal terms. That is, if a measurable factor has a significant association with variation in a given trait—it is a difference that makes a difference—we are meant to envision deliberate alteration of the factor and to expect that that would, all other things being kept the same, result in a change in the trait.

It is easy to find examples to remind us that such a translation from measurable factors to alterable factors is not straightforward. Suicide rates may be higher for men than for women, but no one proposes gender change to a man to reduce his risk of suicide. A positive statistical association between fertilizer level and crop yield does not rule out a negative effect in situations where the fertilizer level is increased to above the usual range. C-reactive protein (CRP) levels in the blood, which rise with inflammation, are associated with an increased incidence of diabetes, hypertension and cardiovascular disease, so we might hypothesize that measures to reduce CRP levels, such as prescribing statins, would reduce the incidence of those medical conditions. Yet we might also hypothesize that some underlying factors lead both to elevation of CRP and increased incidence of the conditions associated with CRP (see section 5, below). In short, it is not surprising that there are active debates about when and how to construe measurable factors in causal terms (Hernan 2002, Freedman 2005).

Suppose that we put aside the temptation to construe measurable factors as alterable or causal factors. Let us see where we get if we return to the rerun quality of additive models, in which they fit well for small differences in some or all of the factors. Because we do not know the underlying causal dynamics, we cannot know what constitutes a *small* difference, and thus when extrapolation will have gone so far that the model's fit to data breaks down. Yet, if we emphasize that the model is being fitted to existing data and that the factors need not be assumed to be causal, we are primed for a key question: *If we modify one or more measured factors, are we modifying the dynamics in a way that disturbs the patterns or associations detected by the analysis of existing data?* In other words, would most policy interventions, if implemented, alter the structure of the relations that produced the phenomena observed and thus the patterns and causal inferences derived from those observations?

For example, high school graduation rates in the United States are associated positively with family income. What would we expect the outcome to be of an annual grant that moves a poor family up to the median income level? Or *outcomes*, given that the supplement might influence more than the risk of the children dropping out of high school. Imagine a scenario in which we are able to identify the ways other aspects of the household and its relations with the local and wider worlds would be influenced by the income supplement. We would need to envision controlling those other aspects so as to leave us only with the extrapolation of the original association between high school graduation rate and family income. Suppose, alternatively, that our only data analysis showed the association in an additive model of high school graduation rate with multiple measured factors. Controlling all the factors other than income would be even more difficult than in the first scenario, given that we do not have an account of the full causal dynamics to guide us. This control would be yet more difficult if there were underlying heterogeneity—dissimilar conjunctions of measurable factors were involved in producing similar responses.

This last remark leads us back to Kendler et al. (2002), which lifts from the full model (Table I.1) a set of separate paths to the outcome to be explained, namely, depression, e.g., "Paths Reflecting a Broad Adversity/Interpersonal Difficulty Pathway to Major Depression." To identify a path on the basis of an additive model is to imply that it is possible to control the factors on all the paths other than one in question. In other words, what gets to be known from a data analysis such as Kendler et al. (2002) is linked not only to the action of excluding therapeutic and policy interventions, but also to the hard-to-envision idea that blocks of measurable factors could be *controlled* while allowing the remaining block to be construed as alterable and causal.

In sum, the reserve professed by Kendler and colleagues about how to interpret their results has been amplified in this section by considering the possibility of underlying heterogeneity and the control of conditions that is implicated in the analysis. Nevertheless, as is to be discussed in Item J, a potentially positive aspect of their research on the association of behavioral traits with an ensemble of measured environmental factors lies in the opening it provides for research that could delve into the underlying heterogeneity.

4. Genome-wide association: Many genetic factors associated with a trait

Genome-Wide Association (GWA) studies have two parts: a) compare a group of individuals—the *case*—who have a disease or other trait with a group of individuals—the *control*—who do not have the trait but match the case in their distribution of other traits; and b) identify places (loci) on the genome that are significantly more frequent in the case compared with the control. (It is also possible through GWA studies to estimate the fraction of the variation in the trait that is associated with the genetic variants; see Item H.6.) The loci in

GWA studies are SNPs (single-nucleotide polymorphisms), which are not the causal genetic factors for the trait, but are simply somewhere close to those factors on the genome. The case and control groups are large (up to around 200,000) and a huge number of SNPs (up to around 1 million) are assayed.

The first successful GWA study was published in 2005. Soon variants were identified that were associated, at least in the defined populations from which the case and control groups were drawn, with increased incidence in diseases such as diabetes, heart disease, and cancers (Khoury et al. 2007). As it has turned out, however, for loci where variants have a statistically significant association with some medically significant trait, that association corresponds to a small increase in incidence of the trait (McCarthy et al. 2008). Expressed in terms of odds, the odds of an individual with the variant having the trait is generally no more than 1.5 times the odds of having it when the variant is not present. Moreover, even when many such associations are considered jointly, most of the variation in the trait remains unaccounted for (Ku et al. 2010; see Item H.6 on so-called *missing heritability*).

The hope had been to expose variants corresponding to a major increase in incidence of the trait, and from that to gain insight into the mechanisms of the disease. Some researchers claim new biomedical insights based on variants corresponding to a small increase (Wheeler and Barroso 2012) or expect the yield from GWA studies to improve once the causally relevant stretches of DNA near the SNPs are identified. Notwithstanding these claims or hopes of new insights, there is active debate on the implications of the difficulty in identifying causally relevant genetic variants through GWA studies (Couzin-Frankel 2010). Some researchers go from the observation that many variants have a small effect (that is, an association with a small increase in incidence of the trait) to a conjecture that future advances in the understanding of a disease will come from finding rare variants (alleles) that have a strong effect (McClellan and King 2010). Genetic heterogeneity is built into this conjecture in the sense that, even if insight about mechanisms emerged from an examination of the strong effect of the rare variant, most of the cases would not be associated with the rare variant. Moreover, the detection and identification of variants is obviously complicated by genetic heterogeneity in its various forms (e.g., mutations in a gene can occur at a variety of points in the gene, the clinical expression of such mutations can vary significantly, and different genetic variants may be expressed as the same clinical entity; see Item G.0).

Heterogeneity is also important in relation to combining results from studies on different populations. Ioannidis et al. (2007) review studies for variants associated with type 2 diabetes in order to show substantial statistical heterogeneity among studies, that is, different variants have statistically significant associations with the trait. The authors make a set of recommendations (see also Gögele et al. 2012), listed below.

a. Undertake *finer mapping* to identify loci close to variants that have effects that are less statistically heterogeneous among studies.

b. Consider *correlated traits* given that many common diseases are associated with others (e.g., as captured by the label *metabolic syndrome*) and investigate whether for any of these the GWA association is more consistent across studies. (The idea here, presumably, is that pathways to the trait originally considered would build on varying combinations of the more consistent correlated traits.)

c. Recognize that diseases "are probably a complex mix of different phenotypes in terms of their molecular pathogenesis," that is, the clinical disease trait results from an *interaction of underlying molecular pathways* and that the specific pathways that go into this interaction "may vary in different subpopulations depending on the presence of other gene variants."

d. Investigate gene-environment interactions and the *variation of environmental exposures* across different populations.

e. Investigate "*genetic heterogeneity*" in effect sizes *across different ethnic backgrounds*."

f. Postulate "population-specific gene-gene *epistatic effects*," but recognize that huge studies would be needed to confirm such interactions.

Most of these recommendations point to possible heterogeneity in the genetic and environmental factors underlying the trait. That is, as noted in Puzzle 2 in Item A, it could be that pairs of SNP variants at a number of loci, say, AAbbccDDee, subject to a sequence of environmental factors, say, FghiJ, are associated, all other things being equal, with the same outcome for the trait as are alleles aabbCCDDEE subject to a sequence of environmental factors FgHiJ. For GWA studies *not* to be disturbed by this possibility, the scope of the results would have to be applied only to the specific population or subpopulation studied. Even under those conditions, however, we would still encounter the same issues about variation within subpopulations and misclassification that were raised for the MAOA case (section 2).

Recall that underlying heterogeneity encompasses both environmental and genetic factors. Nevertheless, several conditions that shape the data analyses almost ensure that the emphasis remains on *genetic* but not environmental heterogeneity. As epidemiologist Frank (2005) has observed, the playing field is not level. Genetic samples are cheap to collect and store and need to be collected only once in a lifetime. Environmental exposures vary over time so that "new samples are needed whenever exposure changes," are difficult to store, and are "getting costlier (as awareness of chemical/physical/ biological complexity increases)." Although some epidemiologists have secured resources to follow small cohorts through time and collect a rich array of data on the individuals (e.g., The Southampton Women's Survey [Inskip et al. 2006]), the major investments are being made in collecting primarily genetic and disease data for large samples (e.g., the UK Biobank). Frank warns that analyses of such data will depend on crude estimates of environmental factors and be subject to large errors, uncertainties, and nonreplicated findings about genetic influences. Yet, if longitudinal data on environmental exposures are sparse, biomedicine will have almost no option but to emphasize the effects of genetic factors and downplay the possibility that the combinations of genetic and environmental factors underlying the trait in question are heterogeneous.

5. Mendelian randomization: Exploration of *environmental* factors for medical traits by controlling a *genetic* factor

Recall the alternative hypotheses arising from the association of C-reactive protein (CRP) levels in the blood with an increased incidence of diabetes, hypertension, and cardiovascular disease (section 3 above). On one hand, measures to reduce CRP levels might reduce the incidence of those medical conditions. Alternatively, some underlying factors might lead both to the elevation of CRP and an increased incidence of the conditions associated with CRP. Ridker, who has shown that inclusion of CRP improves formulas for predicting the incidence of coronary heart disease (Ridker et al. 2007), has been an enthusiastic exponent of the first hypothesis. Although this hypothesis can be tested through clinical trials, a relatively new approach called Mendelian randomization provides a less expensive and quicker test. Mendelian here refers to the existence of variants at some locus on the gene that are associated with increased levels of the risk factor—in this case, CRP. Randomization refers to other risk factors having no association with the presence or absence of that variant (Davey Smith and Ebrahim 2007). It turns out that for four CRP variants, each associated with up to 30% higher CRP levels, no association with coronary heart disease has been found (C Reactive Protein Coronary Heart Disease Genetics Collaboration 2011).

CRP may be thought of as an environmental factor possibly associated with the medical trait in the sense that CRP levels can be altered by medication. Mendelian randomization can also be used to investigate hypotheses about environmental factors in a more conventional sense. Consider the apparent association of moderate alcohol consumption with reduced incidence of coronary heart disease (Davey Smith and Ebrahim 2007, 344ff). Alcohol consumption is also associated with other factors associated with coronary heart disease, such as age, smoking, and an increased level of protective high-density lipoprotein. Moreover, nondrinkers may include people who have stopped drinking because of deterioration in their health. To disentangle these factors, we can look at Japanese populations that have a high frequency of a variant of an enzyme aldehyde dehydrogenase (ALDH2). This variant makes the after-effects of drinking very unpleasant. The average alcohol consumption of these people is low for people who are homozygous for the variant and intermediate for those who are heterozygous. The variant is not associated with—is randomized with respect to—age and smoking, so these factors are not involved in any association that remains between coronary heart disease and the ALDH2 variant—as a proxy for alcohol consumption. It turns out that there is an association between the variant and lower levels of high-density lipoprotein, which supports the causal status of moderate alcohol consumption as a protective factor for coronary heart disease (i.e., associated with a reduced incidence) through the effect of alcohol consumption on levels of high-density lipoprotein. (Davey Smith and Ebrahim 2007 provide a review of other examples and address potential limitations of Mendelian randomization.)

The medical significance of Mendelian randomization lies not in the existence of the genetic variant, which may be quite rare in the population, but in its association with an environmental factor or factors that are modifiable. The control of conditions implicated in the analysis of the data is obvious—it depends on the existence in a population of a Mendelian variant that is independent from the rest of the factors associated with an increased incidence of the trait. If such a variant exists we have as close to a controlled experiment as is possible in situations where multiple environmental factors seem to influence a human trait. Indeed, the control of conditions exceeds that involved when varieties and locations are replicated in agricultural trials.

The control of conditions implicated in the application of the results— assuming the results show that the modifiable factor is causally associated with the trait—is that the factor must be reduced in the population by means that do not exacerbate the other factors associated with the trait. The means of securing that reduction may involve a clinical focus on treating the subset of individuals who have high levels of the factor in question or a public health focus on reducing the average levels of the factor in the population. (The tension between a clinical and public health focus is longstanding in epidemiology and applies not only to the results of Mendelian randomization; Rose 2008 [1992]).

In both the clinical and public health focus, the control of conditions becomes linked to the possibility of heterogeneity. Clinicians who treat patients who have a high level of the factor in question can, at least in principle, pay attention to factors that vary among these individuals. These factors might include levels of other factors associated with the disease (e.g., in the alcohol-coronary heart disease example: smoking, high-density lipoprotein, exercise); responses to medication (or genetic markers for responses to medication); and follow-through on advice given by a clinician. Suppose, however, that the number of individuals having the trait is high outside the subpopulation that has a high level of the factor in question. In that sense there is heterogeneity among those individuals. Clinicians might then want to adjust the criteria for inclusion in the subset to be treated and ask epidemiologists to refine their account of factors associated with the trait. (The refinement may involve factors that are not causal, as appears to be the case with the inclusion of CRP by Ridker et al. 2007 in their improved formula for predicting incidence of coronary heart disease; see above.)

Suppose, as an alternative to the clinical focus, we adopt a public health focus that applies to the whole population. In a population focus, heterogeneity is generally discounted. For example, cigarette taxes and health warning labels may be instituted to reduce the smoking rates even though not all smokers go on to suffer coronary heart disease. However, heterogeneity may be brought back into the picture when the impact of public health measures varies across social groups, e.g., when smoking rates decrease least among people of lower socioeconomic status or have declined, but less so, for working classes (Scollo and Winstanley 2008). In order to achieve control over conditions—smoking rates in this last example—we might need to pay attention to how public health advice is received (Davison et al. 1991) and to other factors that do not apply uniformly—homogeneously—across the population (e.g., workplace stress). In some cases, resistance to the public health measures is fueled by claims that some people—even if only a small subset—are adversely affected by the measure (e.g., low-dose aspirin taken as a prophylaxis against cardiovascular attacks increases nontrivial bleeding events; Seshasai et al. 2012). Pressure then rises to identify subsets that do not show the adverse effect or to convince the population that the benefit-to-risk ratio is sufficiently favorable to go along with the public health measure anyway. In short, the possibility of heterogeneity is relevant in many of the various ways that the control of conditions is implicated in the application of results.

*   *   *   *   *

It is interesting to note that one of the leading exponents of Mendelian randomization is George Davey Smith, a prominent social epidemiologist who became skeptical of three major lines of research in his field: a) the longstanding project of identifying modifiable causes of disease that can be used in efforts to improve population health (Davey Smith 2011; but see Lynch 2007); b) the more recent emphasis on data collection and analysis to separate the effects of diverse biological and social factors that operate on a range of temporal and spatial scales and build up over a person's life course (Davey Smith 2007); and c) the investigation of psychosocial factors associated with disease traits in order to improve prediction of incidence at an individual level (Lynch et al. 2006). To appreciate the difficulty epidemiology has had moving from factors associated with incidence at the population level to improved prediction at an individual level, Davey Smith (2011) invokes the lack of success in the search for systematic aspects of so-called nonshared environmental influences that human behavioral genetics claims overshadow common environmental influences. This part of Davey Smith's argument should be put aside in light of Items H.3 and H.4, which build on Item E and call into question the empirical claim and its interpretation. Even so, we would still be left with the message that epidemiologists should accept the "gloomy prospect" of considerable randomness at the individual level and keep their focus on modifiable causes of disease at the population level. In contrast, the next item suggests that the possible heterogeneity of factors that underlie patterns in observed traits can be studied. Attempting to do so makes sense given the many ways that underlying heterogeneity makes a difference in how heredity and variation are analyzed, interpreted, and acted upon (see Puzzle 2 in Item A, Gap 6 in Item B, and Items C, G, and H).

Item J. Areas of research with potential to allow for the heterogeneity of factors

In which four lines of inquiry are sketched that take us beyond a focus on heredity and variation and into human and social sciences more generally, suggesting ways to allow for the possible heterogeneity of factors of any kind that underlie patterns in observed traits or variables.

In exploring the implications of the possible heterogeneity of underlying factors for research on measurable genetic and environmental factors, Item I also pointed to methodological and other practical difficulties of considering heterogeneity. Item J sketches four lines of inquiry—three actual; one speculative—that have the potential to allow for the possibility of heterogeneity. The intent is not to show that these areas of research overcome the limitations of classical heritability studies discussed in Parts I and II. Instead, each section contributes to posing a puzzle along these lines: What does it take to generate empirically validated models of developmental pathways whose components are heterogeneous and differ among individuals at any one time or over generations? How do we grasp the possibilities and limits of this endeavor?

In addition to the possibility of underlying heterogeneity, this item also addresses the other consideration carried over from Part II, namely, the control of conditions that is implicated in the analysis or, more generally, the actions possible or proposed on the basis of the results.

1. Multivariate additive models of development in education and mental illness

The multivariate additive models of Kendler and colleagues, as described in Item I.3, provide a picture of development that is rich and plausible, but lead to questions about the control of conditions and other actions that are implicated in the analysis. In particular, their research did not include factors that correspond to therapeutic interventions or to social changes that have led to the rising incidence of depression. Ou (2005) also provides a picture of development that is rich and plausible, but the environmental factors in her study have more of an applied-science framing, that is, a closer relation to modification by changes in policy and practice. Her research takes up the following puzzle: Increases in IQ test scores produced by Head Start preschool programs tend to be transient. However, in the long term, through social support systems initiated or enhanced during the Head Start years, the children end up with significantly higher high school graduation rates, employment, and many other socially valued measures (Woodhead 1988). Ou put that conclusion into a quantitative context by finding associations between preschool participation and other measures taken through the course of schooling and development to adulthood. These measures include: basic skills scores at kindergarten; classroom adjustment for ages 9–10; parent involvement for children ages 8–12; abuse/neglect reports for ages 4–12; school quality for children ages 10–14; number of school moves for ages 10–14; commitment to school at age 10 or 15; grade retention through age 15; achievement at age 15; highest grade completed by age 22. Like Kendler et al. (2002), Ou makes caveats, noting that her models are limited by "the correlational nature of the data, possible alternative models, and generalizability" (2005, 604). Even more so, we might add, by the conditions of rerun predictability (Items F, G, and I3).

Kendler et al. (2002) have, however, shown a potential research path that associates behavioral traits with an interaction of many measured environmental factors to delve into the underlying heterogeneity. The researchers tease out from the full model (Table I.1) different kinds of paths to the trait—experiencing an episode of major depression—to be explained. They separate paths reflecting "a broad adversity/interpersonal difficulty pathway to major depression," from those reflecting "a broad externalizing pathway to major depression," and those reflecting "a broad adversity/interpersonal difficulty pathway." The separation of paths might lead us to envision major depression being broken down into multiple traits, e.g., depression following broad adversity or interpersonal difficulty, each trait warranting its own multivariate additive model. That move would represent an attempt to unpack the heterogeneous trait into a set of homogeneous traits. However, we could also see the single trait as an outcome that could be arrived at through heterogeneous pathways or combinations of pathways. Exploration of this second possibility is plausible given the wealth of data behind the published analyses of Ou and Kendler and colleagues. If such exploration proved to be feasible and generated insights, it might also open up possibilities for identifying paths that operate heterogeneously across social groups, across individuals within any social grouping, and, in relation to the Flynn effect (i.e., large increases in average IQ scores from one generation to the next; Puzzle 1 in Item A), heterogeneously across generations.

2. Life course analyses in epidemiology

In a field initiated by the epidemiologist David Barker at the University of Southampton, a large number of researchers have been studying associations between nutritional deficits during critical periods *in utero* and chronic diseases of later life, including heart disease and diabetes. The integration of fetal origins and subsequent influences now takes place under the label of *life course epidemiology* (Kuh and Ben-Shlomo 2004). Gilthorpe and colleagues (among others) have highlighted the statistical challenges in interpreting associations between early life influences and diseases of later life (Head et al. 2005). West and Gilthorpe develop alternative statistical analyses that enable them to characterize different pathways of growth over the life course as separate curves. The separation of these growth curves makes it easier to visualize the possible heterogeneity of factors underlying responses (see also Croudace et al. 2003, DeStavola et al. 2006).

3. Life events and difficulties

Another line of research from England, initiated by the sociologists Brown and Harris in the late 1960s, investigates how severe events and difficulties during people's life course are associated with the onset of mental and physical illnesses (Harris 2000). Brown and Harris have developed a method that is quantitative as well as qualitative, consisting of wide-ranging interviews, ratings of transcripts for the significance of past events in their context (with the rating done blind, that is, without knowledge of whether the person became ill), and statistical analyses. Because what might be recorded as the same event, e.g, death of a spouse, may have very different meanings and significance for different subjects according to the context, life events and difficulties (LEDS) methodology accommodates events with diverse meanings. Reciprocally, apparently heterogeneous events can be subsumed under single factors if the meaning is similar.

For example, in Brown and Harris (1978), depression for working-class women in an area of London is associated with a severe adverse event occurring in the year prior to onset, the lack of a supportive partner, persistently difficult living conditions, and, as a child (under the age of eleven), the loss of, or prolonged separation from, the mother. Each of these factors stands in for a variety of actual events or conditions. Heterogeneity is salient in relation to possible points of intervention or engagement to modify the outcomes (i.e., another instance of actions implicated in the analysis). Let me rehearse the way I would present this last point to students using a reworking by Bowlby (1988) that depicts multiple factors building up over the life course (Figure J.1).

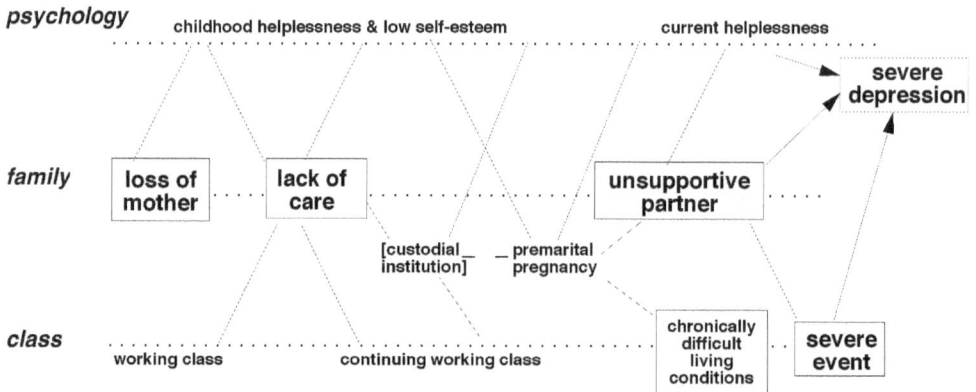

Figure J.1 Depiction by Bowlby (1988, 177) of life development pathways to severe depression identified primarily in the study of working-class women by Brown and Harris (1978). Terms not in boxes, with the exception of the term in brackets, are drawn from a theoretical discussion by Brown and Harris. Bowlby's addition of anxious attachment as a factor in the life course has been omitted. See text for discussion.

Bowlby (1988, 176) notes that the crosscutting dashed lines "indicate the endless ways in which mental state and environment interact," where environment includes the family and socio-economic situations. The lines are dashed, however, to moderate any determinism implied in presenting a smoothed out or averaged schema; the links, while common, do not apply to all women at all times. The links are also contingent on background conditions not shown in the diagram. For example, in a society in which women are expected to be the primary caregivers for children (a background condition), the loss of a mother increases the chances of, or is linked to, the child's lacking consistent, reliable support for at least some period. Given the dominance of men over women and the social ideal of a heterosexual nuclear family, an adolescent girl in a disrupted family or custodial institution would be likely to see a marriage or partnership with a man as a positive alternative. Yet early marriages tend to break up more easily. In a society of restricted class mobility, working-class origins tend to lead to working-class adulthood, in which living conditions are more difficult, especially if a woman has children to look after and provide for on her own. In many such ways these family, class, and psychological strands of the person's life build on each other. Let us also note that, as an unavoidable side effect, the pathways to an individual's depression intersect with and influence other phenomena, such as the state's changing role in providing welfare and custodial institutions, and these other phenomena continue even after the end point, namely, depression, has been arrived at.

Suppose, quite hypothetically, that certain genes, expressed in the body's chemistry, increased a child's susceptibility to helplessness relative to other children in the same circumstances. Suppose also that this inborn biochemistry, or the subsequent biochemical changes corresponding to the helplessness, rendered the child more susceptible to the biochemical shifts that are associated with depression. It is conceivable that early genetic or biochemical diagnosis followed by lifelong treatment with prophylactic antidepressants could reduce the chances of onset of severe depression. This might be true without any other action to ameliorate the effects of the loss of a mother, working-class living conditions, and so on.

There are, however, *many other readily conceivable engagements to reduce the chances of onset of depression.* For example, engagements could include counseling adolescent girls with low self-esteem, quickly acting to ensure a reliable caregiver when a mother dies or is hospitalized, making custodial institutions or foster care arrangements more humane, increasing the availability of contraceptives for adolescents, increasing state support for single mothers, and so on. If the goal is a reduction in depression for working-class women, the unchangeability of the hypothetical inherited genes says nothing about the most effective, economical, or otherwise socially desirable engagement—or combinations of engagements—to pursue.

In short, when teased out by Bowlby into intersecting strands over the life course, the account of Brown and Harris about proximal provoking factors and earlier vulnerability factors, reminds us of a variety—or heterogeneity—of possible points at which engagements could help to modify the outcome. This contrasts with a more limited view of possible action associated with attempts to ascribe causality to underlying genetic factors, albeit modulated, if the nurture: nature ratio is large enough, by environmental factors (Item I).

4. Agent-oriented epidemiology

The methodology of Brown and Harris is labor intensive. Many who have been trained in it have tried to streamline the case-history interview and analysis but eventually shifted away to different lines of research. Suppose, however, that we took as starting points the labor intensity of methodology that exposes the diversity of meanings and the possibility of underlying heterogeneity (i.e., another instance of actions implicated in an analysis). One way forward might be to allow the subjects, living in specific situations that continue to change, to show how they connect knowledge with action. That is, researchers could depart from the traditional emphasis on analyzing data about exposures impinging on subjects, and instead observe communities where people are resilient and reorganize their health, lives, and communities in response to social changes (Sampson et al. 1997). Although the patterns and variation among people revealed in those studies might not extrapolate readily over time, place, and scale, they could provide a point of departure for research and policy engagements in subsequent situations that the researchers study.

Such *agent-oriented* epidemiology would require that researchers be conversant with studies of resilience and reorganization in communities. They would need to train in participant observation and qualitative methods for research on population health changes that arise through grassroots and professional initiatives and then grow into loosely knit social movements, such as groups formed around innovations in short-term therapy for depression (e.g., Griffin and Tyrell 2007, White 2007). In return, they need not feel so gloomy about epidemiologists' limited power to generate reliable and useful associations. Such agent-oriented epidemiology would not be undercut by randomness at the individual level, by heterogeneous pathways and situation-specific meanings for factors, by ongoing change in the causal dynamics that had generated past observations from which statistical associations or patterns had been derived, or by any policy measures they recommend resulting in changes in those dynamics. Dealing with such issues would become a worthwhile challenge, not a quixotic quest.

CODA

At many points during the long history of nature-nurture debates, various scientists and other commentators have asserted that nature versus nurture or genes versus environment is an ill-framed formulation (Paul 1988b). This book clearly adds to that current (as does Tabery 2014). Why then has been hard to just say no to the question *how much is nature, how much is nurture* that persists in popular debate and in announcements about developments in science? The Coda suggests one answer—again offered as an issue to be further expanded on by others. At the same time, this last item complements the conceptual and sociological considerations raised in Item C in taking up the puzzle of how the issue of underlying heterogeneity has been obscured.

*   *   *   *   *

Item K. The persistence of the nature versus nurture formulation viewed in relation to the conflation of family and population

In which it is suggested that the conflation of family and population helps explain why the formulation persists: How much is nature; how much is nurture?

We know that both genes and environment influence outcomes even in the cases in which there is a single gene with a major and direct effect, such as PKU. Recall that in PKU the development of individuals having two copies of a malfunctioning allele for the enzyme phenylalanine hydroxylase (PAH) is extremely impaired by the level of phenylalanine present in normal diets, but much less impaired if a special diet is maintained (Items E.1). Even though in real life this relationship is modulated by a variety of influences (Item I.1), one might still say that genes are primary: Without the genetic condition—the malfunctioning PAH—we do not need to worry about the diet, that is, about the environment. Knowing the genetics—or knowing the biochemistry even before the genetics of the condition is clear—shows us when and how to alter the environment. All kinds of changes in upbringing of individuals with the genetic condition would have no effect. In short, nature interacts with nurture, but it is most important to know about the genetics and associated biochemistry.

If that primacy is the case for PKU, we might well suspect that there are many other genes, perhaps of smaller and less direct effect, for which the same primacy of genetic factors would hold. Why decide in advance that certain traits, such as IQ test scores, are *not* amenable to genetic study? However, let me suggest something other than the possible primacy of genes that might well be involved in the persistence of the nature versus nurture formulation.

Within a family, it is very easy for someone to see that children physically resemble their parents (or the people providing the egg and sperm). Here *resemble* is not simply look like the parents. It means children look more like their parents than any two people randomly picked from the population. In most cases, parents pass environment onto their offspring as well as genes, so we might ask, how important is each? That is hard to say because parents pass on both. So we imagine identical twins raised apart from birth, and ask, Does the one raised at home resemble the parents more than the one raised away? Of course, no offspring fully resembles both parents—and the offspring's sex influences many characteristics (e.g., height, hip width). So, perhaps we should look at resemblance between offspring and the same-sex parent for physical traits (e.g., height) and between offspring and the average of both parents for other traits (e.g., IQ test score). Or perhaps we look for the average of the parents to capture resemblance for physical traits as well and simply expect there to be variation around that average, e.g., female offspring end up on the shorter side of the average and male offspring end up on the taller side.

Once we start talking about variation and averages, we are no longer expecting a clean picture of resemblance in any one family. That means we can shift our sense of resemblance from the family to averages over many offspring-parent pairs. Going back to the identical twins, if we imagine many such pairs of twins, on average does the one raised at home resemble the parents more than the one raised apart? We could also ask, if we have many sets of same-sex, nonidentical twins raised together and many sets of identical twins raised together, does, on average, one identical twin resemble the other twin more than one nonidentical twin resembles the other? (Notice two shifts that simplify the comparison: Looking only at same-sex twins means that we eliminate the complication of differences between an offspring and its other-sex parent. Looking at twins raised together means we do not have to search for the rare cases of twins separated and raised in truly unrelated families.) If the answer in the twin-resemblance study is yes, it seems reasonable to conclude that the identical twins are on average more similar because they share all their genes whereas the nonidentical twins share fewer of their genes. That conclusion does not, however, say that it is the same nature—the same genes—or the same nurture that brings about the resemblance from one pair of twins to the next. (This is the possibility of underlying heterogeneity, raised in Puzzle 2 and discussed especially under Gap 6 in Item B and in Item G.) So then where are we?

It might be useful to consider the results of the study of multiple twins while thinking about your own family. Consider a scenario in which the nonidentical twins are much less similar than identical twins, which means that sharing fewer genes makes a big difference (skipping here the technicalities of getting the number—heritability—that quantifies that result; see Items D and E). If this were the case, for whatever trait we are thinking about, e.g., IQ test score, we might say: "There is nothing I could do as a parent to change the outcome for my offspring. I am not to blame for the outcome other than having passed on my genes." If that conclusion seems justified, we might then reason that the same is true for every other family, and thus society as a whole should not try to change what it is doing because it will not make a difference.

Now flip that scenario. Suppose that the nonidentical twins are just as similar, so that sharing fewer genes makes little difference. What can we do as a parent to make a difference in IQ test scores? (If our offspring are grown, we might ask what *could we have done,* or what could we advise others to do, or what should be done by society at large?) Our study of twins has not shown us what environmental factors have an effect, so we do not know what to change. Moreover, we have to face the possibility of underlying heterogeneity, so that we cannot expect the factors to be the same from one family to the next. We might then just give up on trying to identify those factors.

Notice the asymmetry in these last two paragraphs. Although the possibility of underlying heterogeneity might lead us to give up looking for the relevant environmental factors, it did not lead us to give up on looking for the genetic factors. This is because our reasoning did not lead us to look for those factors at all. We simply concluded that we were not to blame for the outcome in our family and, by extrapolation, society should not try to change what it is doing. This asymmetry should make us suspect that there is a problem in such reasoning in thinking that, because sharing fewer genes makes a big difference, there is nothing a parent can do to change the outcome. Let us tease out what this problem is.

Any set of twins, call it set i, is raised together in family i—each nature i has a nurture i. There are lots of i's. There is nothing in the average over many sets of nature i – nurture i pairs that says there cannot be a nurture j or k or l in which nature i in nurture j or k or l would not be different in interesting ways. Perhaps if we found that identical twins raised apart were just as similar on average as identical twins raised together, we would doubt that such a nurture j or k or l could be found. We might doubt, but we would not be sure. We could end up surprised. After all, Japanese offspring after World War II grew taller on average than their parents, but a comparison of twins in the previous generation would have shown that sharing fewer genes makes a big difference, that is, in technical terms heritability for height would have been high. Yet change turned out to be possible.

Once we entertain the possibility that nature i varies across nurture i, j, k, l,…and that among sets of twins, nature i-nurture i varies, we can ask what genetic factors and what environmental factors are involved and how they act *together* in each combination. The previous asymmetry is gone: For both kinds of factors it is difficult to identify them if you cannot expect them to be common across families.

One response to the difficulty is to put aside the search for common genetic factors and instead use genomics to look for *rare variants* that have a large effect on some biomedical or behavioral condition (McClellan and King 2010). This approach has two justifications:

a. For those individuals carrying the rare variant—or for relatives among whom the variant is common—knowing about the genetics could lead to some constructive responses: genetic counseling about whether to have children at all; screening and selective abortion of fetuses carrying the variant; screening of newborns and other relatives; or investigation of biochemical pathways that might eventually be modified by pharmaceutical treatment.

b. Investigation of biochemical pathways related to the rare variant *might* suggest mechanisms and pharmaceutical treatments to explore that *might* help others who have the condition but lack the particular variant.

For justification b., notice that we have slipped from *family* or other sets of close relatives to a wider *population*. This time the slip results from a hope that the biology is amenable to investigation and intervention that can be extrapolated or generalized. Perhaps that hope will turn out to be realized at least sometimes. If so, the resulting biomedical interventions or treatments would not be based on screening for who has the original genetic variant, but for signs of the biochemical or related conditions that turned out to be amenable to intervention. That kind of screening, of course, dominates Western medicine already. There is nothing especially *genomic* about it.

If we return to the family-based justification for searching for rare variants of large effect (justification a. above), we can envision, because it already happens, that parents whose children possess rare genetic variants of large effect will advocate for more research on the mechanisms and possible treatments. Biomedical researchers have joined forces with such families, in part because the hope of a treatment is important to people who they get to know personally, and in part because researchers benefit to the extent that it boosts the claim that genomics will lead to a revolution in biomedicine (Panofsky 2011). More recently, this claim has mutated: genomics opens up a realm of personalized medicine in which knowledge of one's particular and not-necessarily common genetic profile allows for customized and thus more effective therapies.

There is no denying the emotional power of addressing the needs of families whose children suffer from conditions related to rare variants with a large effect. Notice, however, that addressing these needs is a retreat from the idea of a genomic biomedical revolution where, because all conditions would be linked back to their genetic basis, diseases could be better treated for *everyone* in the population. The idea that research on rare genetic variants that affect some members of some *families* could fuel genomics and a biomedical revolution *for all* speaks again to ease of slipping between family and population.

In the recent tactical regrouping of genomics around the ideal of personalized medicine (Wikipedia n.d., c) the levels of family—or small subpopulation—and population are not well distinguished. The hope is that knowledge of one's particular and not-necessarily common genetic profile allows for customized and thus more effective therapies. However, as discussed in Item I.2, genomics-based pharmaceutical research will generally depend on an advantage *on average* in some given *population* of a treatment—perhaps prophylactic—related to the effect of a genetic variant. Yet there will also surely be variation of effects and responses to drug treatment around that average for the group or subpopulation that have the genetic variant. Given this variation, pressure may emerge to find ways to subdivide the group who has the genetic variant into smaller subcategories so that the treatment can be made to fit each person better. Pharmaceutical companies will not, however, have much incentive to trim the potential market unless there is a strong pushback from people for whom the side effects of treatment far outweigh the benefits or for whom the treatment fails to help. If that pushback happens, negative publicity on those side effects or failures could erode adoption of the drug even among individuals who might have benefited. In sum, the pathway to personalized medicine goes through a phase in which carriers of the genetic variant are treated according to the group to which they belong. The pharmaceutical company may or may not be driven to move beyond that phase to a place where variation within the group or subpopulation is taken into account.

We can be skeptical of the promise represented by genetics or genomics to identify genes for human diseases and behavioral traits without suggesting that genetic factors are uninvolved in the development of traits. The issue here is that these factors are not necessarily the same across all families within a population. Claims that researchers are now finding, or can be expected to find, genes for traits X, Y, and Z abound; caveats about the problem that heterogeneity poses for these claims are rare. Why does this blind spot persist? After all, no technical specialization is needed for someone to understand the possibility of heterogeneity. The reason that the blind spot is hard to dislodge, this item suggests, is the persistent conflation of family and population.

Let me end the Coda with a broad-brush conjecture: The conflation of family and population derives from people being readily able to envision family-level care and support, but not so readily able to envision social-level support on which they can rely. People can see that social institutions do not care well for everyone; some are underserved or left out. Sometimes it seems that institutions "care more about" ensuring their own perpetuation than they do about the wellbeing of individuals. Or, at least, that is common rhetoric. Yet people do have a clear sense of the norm of caring for each other in families even when their own family falls short of that ideal.

The intent of this conjecture is not only to expose the limitations of thinking that researchers will find genes for traits X, Y, and Z, but also to contribute to a sense that social institutions could care in ways beyond treating each individual simply as a member of a population in which all are subject to the same measure. To meet this challenge would be to address the different and overlapping pathways, each consisting of heterogeneous factors that lead to *any* outcome, from our height to our income, health, and happiness. Somehow a caring society has to support each of us in all our heterogeneous developments. How do we, individually and collectively, best contribute to moving toward that goal? To get a handle on that puzzle, readers will certainly have to take the discussion beyond the items in this book. In order to open up spaces for fresh inquiry and action, however, one thing should be clear: the relative weighting of nature and nurture must be put firmly behind us.

Glossary

Items in italics are described elsewhere in the glossary. The index can be consulted for the first mention of the term in the body of the text.

γ: An empirically determined parameter in the gene-free formulation of *quantitative genetics* that takes degree of relatedness into account.

Σ: The sum over possible values of the subscript in the expression that follows.

Additive model: a) Equation that connects the values of the trait for an individual to the summation of several *contributions* (sense b); b) A parallel equation that adds up *variances* related to these contributions.

ANOVA or Analysis of Variance: The *variance* of a trait is divided into parts or *components* corresponding to each term (*contribution* [sense b]) in an *additive model*. For example, for observations of a set of *varieties* raised in a set of *locations*, one part would be the variance of the *variety means*. The other parts would be the location means, the variety-location-interaction means (after subtracting the first two means), and noise, residual or "error" contributions (see equation 1 in Item D).

Behavioral genetics: *Quantitative genetics* applied to behavioral traits, most commonly human behavioral traits.

Cause: A difference that makes a difference. The former difference refers to either a) an intentional modification (e.g., adding fertilizer to an agricultural plot); or b) a distinction between points of data with regard to a measured *factor* that is not, by and large, modifiable (e.g., male versus female sex).

Component: (as used in this book) The parts into which the *ANOVA* divides the *variance* of a trait.

Contribution (as used in this book): a) as colloquially understood, i.e., something given or supplied in common with other contributors; b) a term in an equation that connects the values of the trait for an individual to the summation of several such terms (see *additive model*). Contribution sense b. substitutes for the technical term *effect* because that term is causally ambiguous.

Dizygotic (DZ) twins: Two offspring gestated together from two separately fertilized eggs. Contrast *Monozygotic twins*.

Effect: See *contribution*, sense b.

Environment: a) Set of specific environmental *factors*; or b) Synonym for *location*, a term that can be used without reference to or knowledge of the environmental factors present.

Epigenetics (as the term is used in recent molecular biology): Chemicals from outside the cell can modify the activity of genes for the rest of the organism's life and sometimes even into subsequent generations.

Experimental data: Data derived from explicit manipulation of measured *factors* or conditions.

Factor (as used in this book): Something for which its presence or absence can be observed or its level can be measured—a quality that is emphasized in some places in the text by adding the adjective "measurable." For any given trait, the factors of interest are those associated with changes or variation in the trait's development, but the *causal* quality of a factor is a secondary matter. Measurable genetic factors include the presence or absence of variants (alleles) at a specific place (locus) on a chromosome, repeated DNA sequences, reversed sections of chromosomes, and so on. Measurable environmental factors can range widely, for example, from fertilizer application per hectare of crop to average daily intake of calories to degree of maltreatment that a person experienced as a child.

Genetic (as used in this book): An adjective used in reference to *factors* that are transmitted through the germline from parents to offspring and whose presence or absence can, in principle, be observed.

Genotype: a) The pair of variants (alleles) for a given place or locus on two paired chromosomes; b) A synonym for *variety*, a term that can be used without reference to or knowledge of the genetic factors present.

Genotypic values: See *variety means*.

Heritability: a) The ratio of the *variance* of the *variety means* for a given trait to the overall *variance* of that trait in a population consisting of a specific set of *varieties* raised in some specific *location*(s); b) (recent definition, referred to in this book as new heritability): The fraction of *variance* in a trait associated with variation in Single-Nucleotide Polymorphisms (SNPs) as examined by Genome-Wide Association studies.

Heritability studies: (as used in this book) Research that uses methods of *quantitative genetics* for partitioning variation in a trait into *heritability* and other *components*, such as, the *variance* of the location means.

Heterogeneity: A state in which one kind of thing can be separated into a spectrum, range or mixture of many different kinds. See also *underlying heterogeneity*.

Heterozygous: The two variants that make up a *genotype* (sense a) are different.

Homozygous: The two variants that make up a *genotype* (sense a) are the same.

Interaction: a) A variety-by-location interaction is a *contribution* (sense b) and variety-by-location interaction *variance* is one part in an *Analysis of Variance*. When this variance is high for the trait in question, the ranking of varieties varies across locations and the best variety in one location is not the best in other locations; b) A gene-environment interaction derives from a *regression analysis* involving measured genetic *factors* and measured environmental factors in which included in the *additive model* are additional terms that are the product of a genetic factor and an environmental factor. More generally, interaction means that the quantitative relation between the trait and one of the factors varies according to the measured value of the other factor.

Intraclass Correlation: Ratio of *variances* of the contributions that do not vary within the class (and are thus included in the class averages) to the overall variance. For classes of size two this quantity is mathematically equivalent to the usual linear correlation of the two values in each class or pair, when the order in each pair is arbitrary. Arbitrary ordering would apply if one wanted to know the correlation, for example, of heights in same-sex couples.

Location: The situation or place in which a *variety* is raised, such as a family (for humans) or a specific experimental research station (for agricultural varieties).

Measurable factor: See *factor*.

Monozygotic (MZ) twins: Two offspring gestated together after a single fertilized egg splits and forms two embryos.

Observational data: Data derived on individuals that can be subdivided into relevant categories (e.g., raised in low socioeconomic status) but have not been assigned randomly to be subject to specific conditions (e.g., 50 kg/ha of nitrogen fertilizer) (see *experimental data*).

Path Analysis: Data analysis technique that defines a network of interrelated variables, which may be measurable *factors* or *contributions* estimated in an *Analysis of Variance*, and estimates the relative contribution of each variable to the variation in a focal variable after allowing for the intervening variables. The estimates of relative contributions are called path coefficients. Their reliability depends on the assumptions built into the networks, such as, in *heritability studies*, the similarity of relatives of different degrees and the inclusion in the *additive model* (or exclusion) of coefficients for variety-location *interaction*.

Quantitative genetics: A field in which variation in traits of humans, other animals, or plants is analyzed in ways that take account of the genealogical relatedness of the individuals whose traits are observed. (See also *behavioral genetics* and *heritability studies*.)

Regression analysis: For an equation (often in the form of an *additive model*) that combines a set of measured *factors* (the "independent" variables), the coefficients to each variable are estimated that make the equation predict the trait (the "dependent" variable) better than other values of the coefficients. "Predict best" can be assessed by the lowest residual *variance* (the "least squares") or other criteria.

Replicates: Two or more individuals raised in the same variety-location combination (for heritability studies) or with the same values of the measured factors for *regression analysis*.

Rerun Predictability: The closeness of a match in the following situation: For a set of variety-location combinations trait values are a result of unknown dynamics that include some unsystematic noise. The same variety-location combinations are observed again where the only change in this "rerun" is noise at the same level as the original, but uncorrelated with it. All possible pairs of values are considered in which the first value is from the original situation and the second is from the rerun, where for each pair the original and rerun *variety* is the same (alternatively, the *location* is the same). The closeness of match is assessed by the correlation of the original and rerun values, making use of an *additive model* of *contributions* and assuming that the actual noise is given by the residual or noise *contribution*.

Structural Equation Modeling: A generalization of *path analysis* to include unmeasured "latent" variables that add together several measured *factors* (see *additive model*) in a way that fits the actual data well, not in a predetermined equation.

Trait: Observed characteristic of an individual. (See Item B.1 for explanation of why the term commonly used in *heritability studies*, phenotype, is not used in this book.)

Underlying Heterogeneity: When similar responses of different *varieties* or individual types are observed, but it is not the case that similar conjunctions of genetic and environmental *factors* (or, in epidemiology, risk and protective factors) have been involved in producing those responses (i.e., values for the trait in question).

Variance: The common measure of variation. The variance for the population considers the size of the deviation of an individual's trait from the mean value for the population, squares that, then finds the average over all the individuals. The variety variance considers the size of the deviations of the *variety means* from the overall mean, and so on.

Variety: A group of individuals whose relatedness by genealogy can be characterized, such as offspring of a given pair of parents, or a group of individuals whose mix of genetic *factors* can be replicated, as in an open pollinated plant variety or pure (genetically identical) lines. The term can be used without knowledge of the underlying genetic factors.

Variety mean: The mean or average of the values observed for the trait across all locations in which the *genotype* (sense b) or *variety* is raised, minus the overall mean for the population.

Appendix I. Numerical illustrations of the gene-free formulation of Item D and of the contrasting results under the standard and alternative assumptions of Item E

This appendix provides results from applying the formulas of Items D and E to data sets generated for a range of values of broad-sense heritability and the other fractions of variance. First, data sets of 100 varieties and 100 locations were randomly generated with two MZ twins and two DZ twins for each variety-location combination. Each row of Table 1 corresponds to a pair of V/Y and γ values drawn from the ranges [.2–.8] and [.3–.8], respectively. L, VL, and E were set equal to (1-V)/3. V^- and VL^- were set to $\gamma.V$ and γ .VL, respectively; T and TL were set to $(1- \gamma)V$ and $(1- \gamma)VL$, respectively. Equations 3–7 were used to calculate from the full data set the actual fractions for V, L, VL, and E for MZ and for DZ twins. The average of the two values is given in Table 1. (The actual fractions differ slightly from the pre-set values due to the random generation process.) Equation 12 was used to calculate the actual values for γ. The values for broad-sense heritability and for the fraction of variation due to differences among location averages were estimated first by using the conventional formulas (equations 18 and 19) and then by using the adjusted formulas that factor in γ (equations 15 and 16). The fraction due to noise is 1- the sum of these estimates for broad-sense heritability and the location fraction.

The results in Table 1.1 indicate that: a) the adjusted estimates of broad-sense heritability are close to (V + VL)/Y, and the adjusted values of L/Y are close to the actual values. These results match the theory in the body of the text; b) the conventional formulas yield estimates of the two quantities that differ from the corresponding estimates given by the adjusted formula by a close-to-equal and opposite amount. This amount is close to the value given by the difference between equations 15 and 18, which can be shown to be $(2\gamma -1)$ times the adjusted formula's estimate of V^+/Y.

Table I.I

Actual values				Estimates			
				(conventional formula)		(adjusted formula)	
V/Y	L/Y	VL/Y	γ	Broad-sense heritability	L/Y	Broad-sense heritability	L/Y
.18	.27	.27	.28	.66	.06	.46	.26
.20	.26	.27	.38	.62	.12	.50	.25
.19	.30	.25	.54	.41	.33	.44	.29
.20	.28	.26	.61	.37	.38	.48	.27
.21	.24	.28	.70	.29	.44	.48	.25
.19	.29	.26	.81	.17	.56	.45	.29
.28	.27	.22	.24	.76	.02	.50	.27
.28	.22	.25	.42	.65	.12	.56	.20
.30	.24	.24	.48	.56	.21	.54	.23
.34	.19	.23	.58	.47	.29	.56	.20
.30	.24	.23	.73	.29	.48	.53	.24
.24	.22	.27	.77	.24	.49	.51	.22
.41	.21	.19	.25	.85	-.06	.57	.23
.44	.16	.20	.43	.73	.06	.65	.15
.43	.18	.20	.52	.62	.19	.65	.16
.45	.17	.19	.64	.45	.35	.62	.18
.36	.23	.20	.66	.38	.41	.56	.23
.42	.19	.20	.81	.24	.57	.64	.17
.52	.14	.17	.40	.82	.01	.69	.14
.50	.15	.17	.39	.78	.03	.64	.17
.50	.16	.17	.43	.75	.08	.66	.17
.49	.16	.17	.45	.74	.09	.67	.16
.46	.17	.18	.63	.48	.34	.65	.17
.51	.16	.17	.78	.31	.53	.69	.15
.60	.13	.14	.19	1.18	-.32	.73	.13
.59	.14	.14	.46	.79	.08	.74	.13
.58	.16	.13	.57	.61	.25	.71	.16
.59	.14	.14	.59	.57	.28	.69	.16
.58	.13	.14	.73	.42	.46	.76	.11
.62	.12	.13	.79	.32	.55	.75	.12

.73	.09	.09	.31	1.15	-.24	.83	.08
.70	.08	.11	.43	.90	-.03	.79	.09
.72	.08	.10	.45	.91	-.01	.82	.08
.71	.08	.11	.53	.79	.12	.84	.07
.72	.09	.10	.70	.51	.40	.85	.07
.73	.08	.10	.84	.27	.64	.84	.06
.81	.06	.07	.36	1.13	-.19	.88	.05
.81	.06	.07	.40	1.06	-.12	.88	.05
.81	.06	.07	.41	1.05	-.11	.88	.05
.81	.07	.06	.71	.51	.43	.89	.06
.80	.06	.07	.76	.43	.51	.88	.05
.83	.05	.06	.82	.33	.61	.90	.04

To produce the results given in Table 1.2, 48 MZT pairs, 48 DZT pairs, and 48 UVT pairs were sampled from the full data sets of 10,000 MZ twin pairs and 10,000 DZ twin pairs. (Each row of the table corresponds to a row of Table 1, that is, to a particular pair of V/Y and γ values.) The UVT pairs were sampled from the MZ data by choosing one member of the pairs for two different varieties at a given location. (One sample of pairs for pre-set values of V from .2 to .8 and γ = .5 is downloadable from *http://bit.ly/TwinPairs*.) Both the conventional and adjusted formulas were again used to estimate the values for broad-sense heritability and the fraction of variation due to differences among location averages. After the sampling process was repeated 100 times for each data set, the mean and standard deviation of the estimates were calculated for each quantity.

The results summarized in Table 1.2 have the same features as those in Table 1.1, except that: a) the means of the estimates of broad-sense heritability using the adjusted formula tend to slightly overestimate the actual values for (V + VL)/Y and the means of the corresponding estimates of L/Y tend to underestimate the actual values; and b) the standard deviations of the estimates are substantial, but, in general, are less so for the adjusted than for the conventional formulas.

Table 1.2

Actual Values				Estimates									
				(conventional formula)				(adjusted formula)				γ	
V/Y	L/Y	VL/Y	γ	Broad-sense heritability		L/Y		Broad-sense heritability		L/Y			
				Mean	SD	Mean	SD	Mean	SD	Mean	SD	Mean	SD
.18	.27	.27	.28	.49	.26	.25	.23	.46	.12	.28	.11	.43	.34
.20	.26	.27	.38	.74	.18	.04	.16	.53	.12	.25	.12	.26	.24
.19	.30	.25	.54	.30	.16	.51	.13	.50	.12	.30	.12	.68	.21
.20	.28	.26	.61	.32	.25	.38	.20	.47	.16	.23	.14	.66	.28
.21	.24	.28	.70	.37	.19	.42	.15	.51	.13	.28	.12	.64	.18
.19	.29	.26	.81	.18	.22	.53	.18	.49	.15	.22	.14	.84	.27
.28	.27	.22	.24	.56	.22	.18	.19	.43	.11	.32	.11	.30	.29
.28	.22	.25	.42	.82	.21	-.07	.19	.50	.14	.25	.14	.11	.37
.30	.24	.24	.48	.45	.23	.32	.20	.58	.15	.19	.14	.60	.22
.34	.19	.23	.58	.47	.19	.28	.16	.58	.14	.17	.13	.58	.19
.30	.24	.23	.73	.36	.26	.38	.22	.54	.13	.20	.12	.66	.26
.24	.22	.27	.77	.31	.17	.43	.14	.55	.14	.19	.13	.71	.16
.41	.21	.19	.25	.78	.20	-.02	.17	.52	.15	.24	.14	.17	.36
.44	.16	.20	.43	.43	.17	.38	.15	.62	.14	.18	.14	.65	.15
.43	.18	.20	.52	.75	.19	.04	.17	.68	.13	.11	.13	.43	.19
.45	.17	.19	.64	.37	.21	.42	.18	.65	.14	.14	.14	.71	.17
.36	.23	.20	.66	.32	.22	.43	.18	.54	.15	.21	.14	.71	.21
.42	.19	.20	.81	.17	.17	.60	.13	.61	.14	.16	.13	.86	.15
.52	.14	.17	.40	.99	.23	-.14	.21	.76	.14	.09	.13	.34	.18
.50	.15	.17	.39	.77	.21	.04	.19	.62	.14	.19	.13	.35	.25
.50	.16	.17	.43	.70	.22	.07	.19	.66	.15	.11	.15	.44	.24
.49	.16	.17	.45	.79	.20	.07	.18	.66	.15	.19	.14	.38	.20
.46	.17	.18	.63	.62	.19	.21	.16	.67	.14	.16	.14	.52	.16
.51	.16	.17	.78	.35	.16	.45	.14	.63	.12	.18	.12	.71	.13
.60	.13	.14	.19	1.02	.25	-.18	.24	.67	.14	.17	.13	.22	.23
.59	.14	.14	.46	.93	.14	-.03	.14	.76	.12	.14	.11	.37	.14
.58	.16	.13	.57	.68	.19	.18	.17	.68	.13	.18	.14	.48	.20
.59	.14	.14	.59	.66	.17	.19	.15	.72	.15	.12	.15	.52	.16
.58	.13	.14	.73	.38	.17	.47	.15	.75	.14	.11	.14	.74	.13
.62	.12	.13	.79	.44	.17	.36	.15	.76	.11	.05	.11	.71	.12
.73	.09	.09	.31	.95	.21	-.05	.20	.82	.13	.07	.12	.41	.15
.70	.08	.11	.43	.86	.14	.02	.14	.79	.13	.09	.13	.44	.12
.72	.08	.10	.45	.96	.19	-.05	.18	.80	.13	.11	.13	.39	.16
.71	.08	.11	.53	.74	.17	.15	.16	.87	.13	.02	.13	.57	.12
.72	.09	.10	.70	.49	.12	.43	.11	.86	.14	.06	.14	.71	.08
.73	.08	.10	.84	.26	.09	.64	.08	.85	.15	.05	.15	.84	.06
.81	.06	.07	.36	1.00	.17	-.07	.17	.91	.15	.03	.15	.43	.15
.81	.06	.07	.40	1.09	.20	-.19	.20	.90	.17	-.01	.17	.37	.18
.81	.06	.07	.41	.93	.15	.01	.15	.91	.12	.03	.12	.48	.11
.81	.07	.06	.71	.47	.12	.48	.11	.89	.14	.06	.14	.73	.07
.80	.06	.07	.76	.48	.12	.45	.11	.92	.15	.00	.15	.74	.08
.83	.05	.06	.82	.37	.11	.57	.11	.91	.14	.03	.14	.79	.08

Appendix 2. Ratios of DZ similarity to MZ similarity under the model in which a trait occurs when the dosage from many loci exceeds a threshold (Item E.4.2.b)

The trait occurs when the combined "dosage" from 10 loci exceeds a threshold of 5, where each pair of alleles contributes a full, zero, or half dose according to whether the alleles are, respectively, both the same for one variant, same for the other, or one of each. Dominance is zero for this subset of the full results; location (environment) and noise (error) are not included in the model. The frequency of the first variant for a given locus is randomly chosen from the range in the first two columns. The third column gives the intraclass correlation for DZ twins. Given that the intraclass correlation for MZ twins is 1, the third column also gives the ratio of DZ similarity to MZ similarity. The average of these values is .6.

Table 2.1

Frequency of first kind of allele chosen from range:		Intraclass correlation for DZ twins
Lower limit	Upper limit	
.00	.50	-.01
.49	.51	.78
.38	.63	.61
.25	.75	.53
.13	.88	.73
.00	1.00	.60
.75	.75	.81
.69	.81	.56
.63	.88	.53
.56	.94	.41
.50	1.00	1.00

Appendix 3. Contrasting formulations in a selection of articles that point to the potential importance of interaction variance

This appendix translates the main arguments about the potential importance of interaction variance made by a range of authors into the terms and notation of Part II. This translation makes apparent the differences among the formulations and their differences from the account in this book. Two adjustments should be noted: a) e is used below for environmental factors, not for noise or error; and b) the equations are centered on the average value of the trait over all varieties, locations, and replicates. In other words, in equation 1, subtract m from both sides to produce the deviation of the trait value from the overall mean, then average over the replicates for each variety-location combination and call this average $y'_{ij.}$.

$$y'_{ij.} = \qquad v_i \qquad +l_j \qquad +vl_{ij} \qquad\qquad\qquad (A3.1)$$

I. Layzer (1974)

Main argument: The contribution of variety-location ("genotype-environment") interaction to trait variation consists of a direct interaction contribution and a covariance that is "negligible if, and only if, the genotypic and environmental variables are statistically independent to a high degree of approximation." (Layzer's discussion focuses on variety-location covariance or correlation for human traits.) In mathematical terms:

$$y'_{ii.} = \qquad v_i \qquad +l_i \qquad +vl_{ii} \qquad\qquad\qquad (A3.2)$$

where

$$v_i \quad = \int ... \int \, y\,(g_i,\underline{e})\; p(g_i|\underline{e})\; de_1...de_s \qquad\qquad (A3.3)$$

and g_i is the set of genetic factors present in variety i

$\underline{e}$ is a set of possible values of environmental factors 1,....s

$y\,(g_i,\underline{e})$ is the value of the trait for the set $\underline{e}$

$p(g_i|\underline{e})$ is the conditional probability of the occurrence of for the set $\underline{e}$

$\int ... \int \; ... \; de_1...de_s$ is the integration over all of possible sets of environmental factors 1,....s

Similarly l_i is an integration over all of possible sets of genetic factors. The equation for variances corresponding to A3.2 is

$$Y = V + L + 2\text{ Covariance }(V, L) + 2\text{ Covariance }(V+L, VL) + VL \qquad \text{(A3.4)}$$

which simplifies if covariances are zero to

$$Y = V + L + VL \qquad\qquad\qquad\qquad\qquad\qquad\qquad\qquad\text{(A3.5)}$$

but covariances are zero if and only if $p(g_i|e) = p(g_i)$ and $p(e_i|g) = p(e_i)$.

2. Lewontin (1974a)

Main argument: When the norm of reaction, that is, the response of a variety (or genetic type) to changes in some measurable environmental factor, varies for different varieties in slope or position, this can confound any attempt to extrapolate the relative ranking of varieties observed over part of the range of the environmental factor to the full range. This situation is equivalent to the vl_{ij} values being non-negligible in equation A3.6

$$y'_{ij.} = \qquad v_i \qquad +c.e_j \qquad +vl_{ij} \qquad\qquad\qquad\qquad \text{(A3.6)}$$

where e_j is the average value for the environmental factor in the jth location;

c is a scaling constant; and

vl_{ij} is the additional contribution from the i,j^{th} variety-location combination.

3. Plomin et al. (1977)

Main argument: A proxy for variety-location interaction is constructed, namely, the interaction of two variants of a given variable, e.g., average educational attainment, first averaged for the individual's biological parents (standing in for the variety contribution) and then for the individual's adoptive parents (standing in for the location contribution). Low values of the interaction between the two variants are found for human behavioral traits. In other words, the vl^r_{ij} values are low in the following equation:

$$y'_{ij.} = \qquad v^b_i \qquad +l^a_j \qquad +vl^r_{ij} \qquad\qquad\qquad\qquad \text{(A3.7)}$$

where v^b_i is the average value of y for the biological parents of person i

l^a_j is the average value of y for the adoptive parents of person i raised in family j

vl^r_{ij} is the residual after allowing for the previous two contributions

4. Jacquard (1983)

Main argument: In reality the condition is hardly ever fulfilled that the interaction contributions (estimated by vl_{ij} in equation A3.1) are all zero. Thus heritability in the broad sense (V/Y) has no meaning.

5. Wahlsten (2000)

Main argument: Multiplicative functional relationships can result in a significant interaction variance even if there is no functional interaction term in the functional relationship. Consider the two functional relationships:

$$y'_{ij.} = \quad g_i \quad +e_j \quad +ge_{ij} \tag{A3.8}$$

$$y'_{ij.} = \quad g_i\, e_j \tag{A3.9}$$

where g_i is a functional contribution of the ith variety;

e_j is the functional contribution of the jth location;

ge_{ij} is the functional contribution from the i,j^{th} variety-location combination.

If the ge_{ij} values in equation A3.8 are negligible, the values of vl_{ij} in equation A3.1 will also be negligible. However, even without any ge_{ij} values in A 3.9, statistically significant values of vl_{ij} can arise.

Summary

In Part II, the linear model underlying the partitioning of variation in the trait connects values of the trait for an individual to the summation of variety, location, variety-location interaction, and noise contributions (e.g., equation 1). It does so without reference to measurable genetic or environmental factors (unlike #1, 5), their functional relationships (unlike #5), probabilities of their occurrence (unlike #1), or a gradient underlying the differences in the variety or location contributions (unlike #2). Interaction variance can be estimated (i.e., no proxy is needed; unlike #3) provided data are available from the appropriate classes of defined degrees of relatedness. In that case, broad-sense heritability can be estimated (unlike #4).

Several of the authors also make arguments about the potential importance of variety-location ("genotype-environment") correlation (or covariance). In Part II, lack of such correlations is assumed in order to keep the focus on partitioning of variance under three standard assumptions and their alternatives. As noted in Item D.4, whether lack of such correlations and other conditions can be met in human situations is a matter of ongoing controversy.

Appendix 4. Responses and counter-responses to critiques of heritability studies

Table 4.1 Summary of responses and counter-responses to critiques of heritability studies*

Response	Counter-Response
There are many definitions or formulas for heritability other than the ratio of the variance among variety means (genotypic values) to the variance of the trait across the whole data set.	A sound justification for using other definitions of *heritability* is that they approximate either the original definition or the predictions that definition provides of advance under selection.
Narrow-sense heritability is favored over the broad-sense heritability above.	Narrow-sense heritability can only be defined when quantitative genetics employs models of theoretical, idealized genes with simple Mendelian inheritance and direct contributions to the trait. However, a gene-free analysis of trait variation must also be possible (Item D).
Genotype-environment interaction variance (here: variety-location interaction variance) has been shown to be negligible for humans.	From Item E.4.3: "Plomin et al. (1977), which is often cited in the context of low interaction variance for humans, considers only a proxy for variety-location interaction. For some trait, e.g., educational attainment, the statistical interaction of the average for biological parents with the average for adoptive parents is calculated. How well such proxy results reflect actual variety-location interaction is, however, hard to assess in the absence of studies for a range of human traits in which the classes of data are collected that allow VL/Y to be separated out from V^+/Y."

Response	Counter-Response
Gene-environment interaction is not assumed to be negligible. Indeed, there is an institutionalized field of research on the topic.	The use of the term *interaction* in this response refers to the statistical interaction between measured genetic factors and measured environmental factors. This is conceptually and empirically distinct from variety-location (genotype-environment) interaction, in which a high value means that, for the trait in question, the ranking of varieties varies across locations.
The intraclass correlation formulas in heritability studies have been superseded by calculations that use Structural Equation Modeling.	SEM shares the features that render unreliable estimates made using intraclass correlation formulas (see Item E, especially assumptions 2 and 3).
"Research into the genetics of complex traits has moved from the estimation of genetic variance in populations to the detection and identification of variants that are associated with or directly cause variation" (Visscher 2007). The abundance of genomic data changes the landscape completely.	Classical quantitative genetic heritability studies are still used, for example, to indicate that "the trait [is] a potentially worthwhile candidate for molecular research" to identify the specific genetic factors involved (Nuffield Council on Bioethics 2002, chap. 11). In turn, the difficulty of identifying such genetic factors has led to concerns about "missing heritability" (e.g., Manolio et al. 2009).
It is past time to move beyond debates on the existence of genetic influences.	The problems conveyed in Parts I and II are worth recognizing: they call most estimates and interpretations from human quantitative genetics into question. The possibility of underlying heterogeneity that arises from this more critical account helps clarify the challenging paths ahead.
Genetically informed social science should not be compared with agricultural trials, but be seen as an improvement on social science that is not genetically informed.	Same as above, where the challenging paths extend to bringing genetic factors into social science in a reliable manner (see Part III).

Response	Counter-Response
Actual data should be analyzed to show that alternative assumptions make a difference.	The implications drawn at a theoretical level are confirmed by analysis of simulated data (Item E and Appendices 1 and 2). There are no grounds to suspect that, if actual data were available of the kind needed for estimation under the alternative assumptions, these implications would no longer hold. Even if there turn out to be cases in which differences between the estimates are negligible, the existence of alternative assumptions warrants recognition.
"It is one thing to criticize the methodology of specific studies. It is quite another to suggest... that we reject the results of an entire field of scientific inquiry... It is highly unlikely that modern psychiatric genetics will be judged by future historians of science to be in such company [as astrology and alchemy]." (Kendler 2005, 10)	From end of Item C: "Perhaps human quantitative genetics could be viewed, *contra* Kendler, as akin to alchemy, a field of inquiry that provided observations, questions, tools, debates, careers, and institutions which modern chemistry built on, but ultimately had to break away from to make further progress..."
Some assumption has to be made that connects similarity in traits for relatives to the fraction shared by the relatives of all the genes that vary in the population. Direct proportionality is as good as any other assumption.	No assumption has to be made because, in a gene-free analysis of trait variation, the empirical estimation of a parameter to take degree of relatedness into account is possible (Item D). In any case, estimates should indicate the sensitivity to any such assumption.
It is well known that high heritability does not show where to look for the variants.	It is not widely acknowledged that high heritability is not a good indicator of which traits are worthwhile candidates for molecular research.

Response	Counter-Response
Human quantitative genetics is subject to various well-recognized, albeit contested objections. For example: a) Has zygosity of twins been correctly ascertained and representatively sampled? b) Is the treatment or experience of the twins or unrelated individuals within a location unaffected by whether they are monozygotic twins, dizygotic twins, or unrelated? (Item D.4)	A focus on such technical issues runs the risk of missing or even reinforcing the more fundamental assumptions identified in Item E. For example, issue b. emerges from discussions that take heritability as a measure of the influence of genes and the percentage not included in heritability as a measure of the influence of environmental factors. Examining how well twins share specific environmental factors risks obscuring the difficult-to-bridge gap between the statistical patterns of heritability studies and measurable underlying factors (see Gaps 2 and 3 in Item B and Puzzle 2 in Item A).

* The responses have been drawn in large part from reviewers of different versions of what was eventually published as Taylor (2012), on which Items D and E are based.

Bibliography

Alós, T., Y. Bel, et al. (1993). "Improved identification of heterozygotes for phenylketonuria using blood neopterin and biopterin." *Journal of Inherited Metabolic Disease* 16(2): 457-464.

Betzig, L., M. B. Mulder, et al., Eds. (1988). *Human Reproductive Behaviour: A Darwinian Perspective*. Cambridge: Cambridge University Press.

Bouchard, T. and M. McGue (1981). "Familial studies of intelligence: a review." *Science* 212(4498): 1055-1059.

Bowlby, J. (1988). *A Secure Base*. New York: Basic Books.

Brown, G. W. and T. Harris (1978). *Social Origins of Depression*. New York: The Free Press.

Bruder, C. E. G., A. Piotrowski, et al. (2008). "Phenotypically concordant and discordant monozygotic twins display different DNA copy-number-variation profiles." *American Journal of Human Genetics* 82: 763-771.

Brunner, H. G., M. Nelen, et al. (1993). "Abnormal behavior associated with a point mutation in the structural gene for monoamine oxidase A." *Science* 262: 578-580.

Burks, B. S. (1928). "The relative influence of nature and nurture upon mental development: a comparative study of foster parent-foster child resemblance and true parent-true child resemblance." *The Twenty-Seventh Yearbook of the National Society for the Study of Education* 27: 219-316.

Byth, D. E., R. L. Eisemann, et al. (1976). "Two-way pattern analysis of a large data set to evaluate genotypic adaptation." *Heredity* 37(2): 215-230.

C Reactive Protein Coronary Heart Disease Genetics Collaboration (2011). "Association between C reactive protein and coronary heart disease: Mendelian randomisation analysis based on individual participant data." *British Medical Journal* 342(15 February): d548.

Caspi, A., J. McClay, et al. (2002). "Role of genotype in the cycle of violence in maltreated children." *Science* 297(5582): 851-854.

Caspi, A., K. Sugden, et al. (2003). "Influence of life stress on depression: moderation by a polymorphism in the 5-HTT gene." *Science* 301(5631): 386-389.

Couzin-Frankel, J. (2010). "Major heart disease genes prove elusive." *Science* 328(5983): 1220-1221.

Croudace, T. J., M.-R. Jarvelin, et al. (2003). "Developmental typology of trajectories to nighttime bladder control: epidemiologic application of longitudinal latent class analysis." *American Journal of Epidemiology* 157: 834–842.

Davey Smith, G. (2007). "Lifecourse epidemiology of disease: a tractable problem?" *International Journal of Epidemiology* 36(3): 479-480.

Davey Smith, G. (2011). "Epidemiology, epigenetics and the 'Gloomy Prospect': Embracing randomness in population health research and practice." *International Journal of Epidemiology* 40: 537-562.

Davey Smith, G. and S. Ebrahim (2007). "Mendelian randomization: Genetic variants as instruments for strengthening causal influences in observational studies. Pp. 336-366 in *Biosocial Surveys*. M. Weinstein, J. W. Vaupel and K. W. Wachter (Eds.) Washington, DC: National Academies Press.

Davison, C., G. Davey Smith, et al. (1991). "Lay epidemiology and the prevention paradox: The implications of coronary candidacy for health education." *Sociology of Health and Illness* 13(1): 1-19.

De Stavola, B. L., I. dos Santos Silva, et al. (2004). "Childhood growth and breast cancer." *American Journal of Epidemiology* 159(7): 671-682.

Dickens, W. T. and J. R. Flynn (2001). "Heritability estimates versus large environmental effects: The IQ paradox resolved." *Psychological Review* 108(2): 346-369.

Downes, S. M. (2004). "Heredity and heritability." In *The Stanford Encyclopedia of Philosophy*. E. N. Zalta (Ed.) *http://plato.stanford.edu/entries/heredity/* (viewed 7 August 2014).

Falconer, D. S. and T. F. C. Mackay (1996). *Introduction to Quantitative Genetics*. Harlow: Longman.

Fausto-Sterling, A. (2014). "Letting go of normal." *Boston Review* (March/April), *http://www.bostonreview.net/wonders/fausto-sterling-motor-development* (viewed 21 June 2014).

Feng, R., G. Zhou, et al. (2009). "Analysis of twin data using SAS" Biometrics 65: 584–589.

Flynn, J. R. (1994). "IQ gains over time." Pp. 617-623 in *Encyclopedia of Human Intelligence*. R. J. Sternberg (Ed.) New York: Macmillan.

Flynn, J. R. (2000). *How to Defend Humane Ideals: Substitutes for Objectivity*. Lincoln, NE: University of Nebraska Press.

Frank, J. (2005). "A tale of (more than?) two cohorts—from Canada." *3rd International Conference on Developmental Origins of Health and Disease*.

Freedman, D. A. (2005). "Linear statistical models for causation: A critical review" in *Encyclopedia of Statistics in the Behavioral Sciences*. B. Everitt and D. Howell (Eds.) Hoboken, NJ: Wiley.

Fryer, R. and S. Levitt (2004). "Understanding the black-white test score gap in the first two years of school." *The Review of Economics and Statistics* 86(2): 447-464.

Gatzke-Kopp, L. M., M. T. Greenberg, et al. (2012). "Aggression as an equifinal outcome of distinct neurocognitive and neuroaffective processes." *Development and Psychopathology* 24(Special Issue" 03): 985-1002.

Gilbert, S. (2013). *Developmental Biology*. Sunderland, MA: Sinauer.

Gögele, M., C. Minelli, et al. (2012). "Methods for meta-analyses of genome-wide association studies: critical assessment of empirical evidence." *American Journal of Epidemiology* 175(8): 739-749.

Griffin, J. and I. Tyrrell, Eds. (2007). *An Idea in Practice: Using the Human Givens Approach*. Chalvington, UK: Human Givens Publishing.

Griffiths, P. E. and J. Tabery (2008). "Behavioral genetics and development: Historical and conceptual causes of controversy." *New Ideas in Psychology* 26(3): 332-352.

Harris, T., Ed. (2000). *Where Inner and Outer Worlds Meet*. London: Routledge.

Head, R. F., G. T. F. Ellison, et al., Eds. (2005). *Statistical challenges facing the foetal origins of adult disease hypothesis*. 3rd International Congress on the Developmental Origins of Health and Disease, Toronto.

Hernan, M. A. (2002). "Causal knowledge as a preequisite for confounding evaluation: an application to birth defects epidemiology." *American Journal of Epidemiology* 155(2): 176-184.

Holland, J. B., W. E. Nyquist, et al. (2003). "Estimating and interpreting heritability for plant breeding: An update." *Plant Breeding Reviews* 22: 9-112.

Howell, D. C. (2002). "Intraclass correlation: for unordered pairs." *http://www.uvm.edu/~dhowell/StatPages/More_Stuff/icc/icc.html* (viewed 21 June 2014).

Hudson, C. G. (2005). "Socioeconomic status and mental illness: tests of the social causation and selection hypotheses." *American Journal of Orthopsychiatry* 75(1): 3-18.

Inskip, H. M., K. M. Godfrey, et al. (2006). "Cohort profile: The Southampton Women's Survey." *International Journal of Epidemiology* 35(1): 42-48.

Ioannidis, J. P. A., N. A. Patsopoulos, et al. (2007). "Heterogeneity in meta-analyses of genome-wide association investigations." *PLoS ONE* 2(9): e841.

Jacquard, A. (1983). "Heritability: One word, three concepts." *Biometrics* 39: 465-477.

Jencks, C. and M. Phillips, Eds. (1998). *The Black-White Test Score Gap*. Washington, DC: Brookings Institution Press.

Jensen, A. R. (1969). "How much can we boost IQ and scholastic achievement?" *Harvard Educational Review* 39: 1-123.

Jensen, A. R. (1970). "Race and the genetics of intelligence: A reply to Lewontin." *Bulletin of the Atomic Scientists* 26: 17-23.

Kaplan, J. M. (2000). *The Limits and Lies of Human Genetic Research*. New York: Routledge.

Keller, E. F. (2010). *The Mirage of a Space between Nature and Nurture*. Durham, NC: Duke University Press.

Kendler, K. S. (2005). "Reply to J. Joseph, research paradigms of psychiatric genetics." *American Journal of Psychiatry* 162: 1985-1986.

Kendler, K. S. and J. H. Baker (2007). "Genetic influences on measures of the environment: a systematic review." *Psychological Medicine* 37(5): 615-626.

Kendler, K. S., C. O. Gardner, et al. (2002). "Towards a comprehensive developmental model for major depression in women." *American Journal of Psychiatry* 159: 1133-1145.

Kendler, K. S. and C. A. Prescott (2006). *Genes, Environment, and Psychopathology: Understanding the Causes of Psychiatric and Substance Abuse Disorders*. New York: The Guilford Press.

Khoury, M. J., J. Little, et al. (2007). "On the synthesis and interpretation of consistent but weak gene-disease associations in the era of genome-wide association studies." *International Journal of Epidemiology* 36: 439-445.

Ku, C. S., E. Y. Loy, et al. (2010). "The pursuit of genome-wide association studies: Where are we now?" *Journal of Human Genetics* 55(April): 195-206.

Kuh, D. and Y. Ben-Shlomo, Eds. (2004). *A Life Course Approach to Chronic Disease Epidemiology*. Oxford: Oxford University Press.

Layzer, D. (1974). "Heritability analyses of IQ scores: Science or numerology?" *Science* 183(4131): 1259 - 1266.

Lewontin, R. C. (1970a). "Race and intelligence." *Bulletin of the Atomic Scientists* 26: 2-8.

Lewontin, R. C. (1970b). "Further remarks on race and the genetics of intelligence." *Bulletin of the Atomic Scientists*: 23-25.

Lewontin, R. C. (1974a). "The analysis of variance and the analysis of causes." *American Journal of Human Genetics* 26: 400-411.

Lewontin, R. C. (1974b). *The Genetic Basis of Evolutionary Change.* New York: Columbia University Press.

Lewontin, R. C. (1982). *Human Diversity.* New York: Freeman Press.

Lindman, H. R. (1992). *Analysis of Variance in Experimental Design.* New York: Springer-Verlag.

Lush, J. L. (1947). "Family merit and individual merit as bases for selection." *American Naturalist* 81: 241-261; 362-379.

Lynch, J., G. Davey Smith, et al. (2006). "Explaining the social gradient in coronary heart disease: comparing relative and absolute risk approaches." *Journal of Epidemiology and Community Health* 60: 436-441

Lynch, M. and B. Walsh (1998). *Genetics and Analysis of Quantitative Traits.* Sunderland, MA: Sinauer.

Majumder, P. P. and S. Ghosh (2005). "Mapping quantitative trait loci in humans: achievements and limitations." *Journal of Clinical Investigation* 115(6): 1419-1424.

Manolio, T. A., Collins, F. S., et al. (2009). "Finding the missing heritability of complex diseases." *Nature, 461*: 747-753.

McCarthy, M. I., G. R. Abecasis, et al. (2008). "Genome-wide association studies for complex traits: consensus, uncertainty and challenges." *Nature Reviews Genetics* 9(May): 356-369.

McLaughlin, P. (1998). "Rethinking the Agrarian Question: The limits of essentialism and the promise of evolutionism." *Human Ecology Review* 5(2): 25-39.

McClellan, J. and M.-C. King (2010). "Genetic heterogeneity in human disease." *Cell* 141: 210-217.

Michala, L., D. Goswami, et al. (2008). "Swyer syndrome: Presentation and outcomes." *BJOG: An International Journal of Obstetrics & Gynaecology* 115(6): 737-741.

Miele, F. (2002). *Intelligence, Race, and Genetics: Conversations with Arthur Jensen.* Boulder, CO: Westview Press.

Moffitt, T. E., A. Caspi, et al. (2005). "Strategy for investigating interactions between measured genes and measured environments." *Archives of General Psychiatry* 62(5): 473-481.

Morris, C., A. Shen, et al. (2007). "Deconstructing violence." *GeneWatch* 20(2):

3-10.

Neisser, U., G. Boodoo, et al. (1996). "Intelligence: Knowns and unknowns." *American Psychologist* 51: 77-101.

Nisbett, R. E. (1998). "Race, genetics, and IQ." Pp. 86-102 in *The Black-White Test Score Gap*. C. Jencks and M. Phillips (Eds.) Washington, DC: Brookings Institution Press.

Nisbett, R. E., J. Aronson, et al. (2012). "Intelligence: New findings and theoretical developments." *American Psychologist* 67(2): 130-159.

Nuffield Council on Bioethics (2002). *Genetics and Human Behavior: The Ethical Context. http://www.nuffieldbioethics.org* (viewed 22 June 2007).

Otto, S. P., F. B. Christiansen, et al. (1995). *Genetic and cultural inheritance of continuous traits*. Stanford University Morrison Institute for Population and Resource Studies Working Paper Series No. 64, *http://hsblogs.stanford. edu/morrison/morrison-institute-working-papers-pdf* (viewed 21 June 2014).

Ou, S.-R. (2005). "Pathways of long-term effects of an early intervention program on educational attainment: Findings from the Chicago longitudinal study." *Applied Developmental Psychology* 26: 478-611.

Panofsky, A. (2011). "Generating sociability to drive science: Patient advocacy organizations and genetics research." *Social Studies of Science* 41(1): 31–57.

Panofsky, A. (2014). *Misbehaving Science: Controversy and the Development of Behavior Genetics*. Chicago: University of Chicago Press.

Parens, E. A. and N. Chapman, Eds. (2006). *Wrestling with Behavioral Genetics: Science, Ethics, and Public Conversation*. Baltimore: Johns Hopkins University Press.

Paul, D. B. (1998). "A debate that refuses to die." Pp. 81-93 in *The Politics Of Heredity: Essays on Eugenics, Biomedicine, and The Nature-Nurture Debate*. Albany: SUNY Press.

Paul, D. B. (2000). "A double-edged sword." Nature 405: 515.

Paul, D. B. and J. P. Brosco (2013). *The PKU Paradox: A Short History of a Genetic Disease*. Baltimore: Johns Hopkins University.

Pinker, S. (2004). "Why nature and nurture won't go away." *Daedalus* (Fall): 1-13.

Plomin, R. (1999). "Genetics and general cognitive ability." *Nature* 402: C25-C29.

Plomin, R. (2011). "Commentary: Why are children in the same family so different? Non-shared environment three decades later." *International Journal of Epidemiology* 40(3): 582-592.

Plomin, R. and K. Asbury (2006). "Nature and nurture: genetic and environmental influences on behavior." *The Annals of the American Academy of Political and Social Science* 600(1): 86-98.

Plomin, R., J. C. DeFries, et al. (1977). "Genotype-environment interaction correlation in analysis of human behavior." *Psychological Bulletin* 84(2): 309-322.

Plomin, R., J. C. Defries, et al. (1997). *Behavioral Genetics*. New York: Freeman.

Principe, L. M. (2007). "A revolution nobody noticed? Changes in early eighteenth-century chymistry." Pp. 1-22 in *New Narratives in Eighteenth-Century Chemistry*. L. M. Principe (Ed.) Dordrecht: Springer

Principe, L. M. and W. R. Newman (2001). "Some problems in the historiography of alchemy." Pp. 385-434 in *Secrets of Nature: Astrology and Alchemy in Early Modern Europe*. W. R. Newman and A. Grafton (Eds.) Cambridge, MA: MIT Press.

Richardson, K. and S. Norgate (2005). "The equal environments assumption of classical twin studies may not hold." *British Journal Educational Psychology* 75(3): 339-350.

Ridker, P. M., J. E. Buring, et al. (2007). "Development and validation of improved algorithms for the assessment of global cardiovascular risk in women: The Reynolds Risk Score." *Journal of the American Medical Association* 297: 611-619.

Rijsdijk, F. V. and P. C. Sham (2002). "Analytic approaches to twin data using structural equation models." *Briefings In Bioinformatics* 3(2): 119–133.

Risch, N., R. Herrell, et al. (2009). "Interaction between the serotonin transporter gene (5-HTTLPR), stressful life events, and risk of depression: A meta-analysis." *Journal of the American Medical Association* 301(23): 2462-2471.

Rose, G. (2008 [1992]). *Rose's Strategy of Preventive Medicine*. Oxford: Oxford University Press.

Rutter, M. (2002). "Nature, nurture, and development: From evangelism through science toward policy and practice." *Child Development* 73: 1-21.

Sampson, R. J., S. W. Raudenbush, et al. (1997). "Neighborhoods and violent crime: A multilevel study of collective efficacy." *Science* 277: 918-924.

Sapp, J. (1983). "The struggle for authority in the field of heredity, 1900-1932: New perspectives on the rise of genetics." *Journal of the History of Biology*

16(3): 311-342.

Sarkar, S. (1998). *Genetics and Reductionism*. Cambridge: Cambridge University Press.

Scarr, S. and K. McCartney (1983). "How people make their own environments: A theory of genotype-environment effects." *Child Development* 54: 424-435.

Schiff, M. and R. C. Lewontin (1986). *Education and Class: The irrelevance of IQ genetic studies*. New York: Oxford University Press.

Scollo, M. M. and M. H. Winstanley (2008). "1.7 Trends in the prevalence of smoking by socioeconomic status." In *Tobacco in Australia: Facts and Issues*. M. M. Scollo and M. H. Winstanley (Eds.) Melbourne: Cancer Council Victoria.

Sesardic, N. (2005). *Making Sense of Heritability*. Cambridge: Cambridge University Press.

Seshasai, S. R. K., S. Wijesuriya, et al. (2012). "Effect of aspirin on vascular and nonvascular outcomes: Meta-analysis of randomized controlled trials." *Archives of Internal Medicine* 172: 209-216.

Shim, J. K., K. W. Darling, et al. (2014). "Homogeneity and heterogeneity as situational properties: Producing—and moving beyond?—race in post-genomic science." *Social Studies of Science*, *http://sss.sagepub.com/content/early/2014/05/13/0306312714531522* (viewed 21 June 2014).

Stewart, J. (1979). "Scientific findings that look awkward for socialists: How are we to respond?" *Radical Science* 8: 121-123.

Sweetman, L. (2001). "Newborn screening by tandem mass spectrometry: Gaining experience." *Clinical Chemistry* 47(11): 1937-1938.

Tabery, J. (2009). "Making sense of the nature–nurture debate, Review of Neven Sesardic (2005), Making Sense of Heritability." *Biology and Philosophy* 24: 711–723.

Tabery, J. (2014). *Beyond Versus: The Struggle to Understand The Interaction of Nature and Nurture*. Cambridge, MA: MIT Press.

Taylor, P. J. (2001). "Distributed agency within intersecting ecological, social, and scientific processes". Pp. 313-332 in *Cycles of Contingency: Developmental Systems and Evolution*. S. Oyama, P. Griffiths and R. Gray (Eds.) Cambridge, MA: MIT Press.

Taylor, P. J. (2004). "What can we do? -- Moving debates over genetic determinism in new directions." *Science as Culture* 13(3): 331-355.

Taylor, P. J. (2005). *Unruly Complexity: Ecology, Interpretation, Engagement.* Chicago: University of Chicago Press.

Taylor, P. J. (2006a). "Heritability and heterogeneity: On the limited relevance of heritability in investigating genetic and environmental factors." *Biological Theory* 1(2): 150-164.

Taylor, P. J. (2006b). "Heritability and heterogeneity: On the irrelevance of heritability in explaining differences between means for different human groups or generations." *Biological Theory* 1(4): 392-401.

Taylor, P. J. (2007). "The Unreliability of High Human Heritability Estimates and Small Shared Effects of Growing Up in the Same Family." *Biological Theory* 2(4): 387-397.

Taylor, P. J. (2008a). "The under-recognized implications of heterogeneity: Opportunities for fresh views on scientific, philosophical, and social debates about heritability." *History and Philosophy of the Life Sciences* 30: 431-456.

Taylor, P. J. (2008b). "Underlying heterogeneity: A problem for biological, philosophical, and other analyses of heritability?" *Biology and Philosophy* 23(4): 587-589.

Taylor, P. J. (2008c). "Developing critical thinking is like a journey." Pp. 155-169 in *Teachers and Teaching Strategies, Problems and Innovations.* G. F. Ollington (Ed.). Hauppauge, NY: Nova Science Publishers.

Taylor, P. J. (2009a). "Nothing reliable about genes or environment: New perspectives on analysis of similarity among relatives in light of the possibility of underlying heterogeneity." *Studies in History and Philosophy of Biological and Biomedical Sciences* 40(3): 210-220.

Taylor, P. J. (2009b). "Infrastructure and scaffolding: Interpretation and change of research involving human genetic information." *Science as Culture* 18(4): 435-459.

Taylor, P. J. (2010). "Three puzzles and eight gaps: What heritability studies and critical commentaries have not paid enough attention to." *Biology & Philosophy* 25(1): 1-31.

Taylor, P. J. (2011). "What to do if we think that researchers have overlooked a significant issue for a long time? The case of quantitative genetics and underlying heterogeneity." Paper presented to the History and Philosophy of Science Department, Indiana University (September 23).

Taylor, P. J. (2012). "A gene-free formulation of classical quantitative genetics used to examine results and interpretations under three standard assumptions." *Acta Biotheoretica* 60(4): 357-378.

Turkheimer, E. (2000). "Three laws of behavior genetics and what they mean." *Current Directions in Psychological Science* 9(5): 160-164.

Turkheimer, E. (2004). "Spinach and ice cream: Why social science is so difficult." Pp. 161-189 in *Behavior genetics principles: Perspectives in development, personality, and psychopathology*. L. DiLalla (Ed.) Washington, DC: American Psychological Association.

Turkheimer, E. (2008). "A better way to use twins for developmental research." *LIFE Newsletter(Max Planck Institute for Human Development)*(Spring): 2-5.

Turkheimer, E., A. Haley, et al. (2003). "Socioeconomic status modifies heritability of IQ in young children." *Psychological Science* 16(6): 623-628.

Turkheimer, E. and M. Waldron (2000). "Nonshared environment: A theoretical, methodological, and quantitative review." *Psychological Bulletin* 126(1): 78-108.

Visscher, P. M., S. Macgregor, et al. (2007). "Genome partitioning of genetic variation for height from 11,214 sibling pairs." *American Journal of Human Genetics* 81: 1104-1110.

Visscher, P. M., S. E. Medland, et al. (2006). "Assumption-free estimation of heritability from genome-wide identity-by-descent sharing between full siblings." *PLoS Genetics* 2(3).

Wheeler, E. and I. Barroso (2011). "Genome-wide association studies and type 2 diabetes." *Briefings in Functional Genomics* 10 (2): 52-60.

White, M. (2007). *Maps of Narrative Practice*. New York: Norton.

Wikipedia (n.d., a) "Heritability." *http://en.wikipedia.org/wiki/Heritability* (viewed 14 Mar 2008).

Wikipedia (n.d., b) "Gene-environment correlation." *http://en.wikipedia. org/wiki/Gene-environment_correlation* (viewed 14 Mar 2008).

Wikipedia (n.d., c) "Personalized medicine." *http://en.wikipedia.org/wiki/ Personalized_medicine* (viewed 29 Jun 2013).

Williams, R. (1980). "Ideas of nature." Pp. 67-85 in *Problems in Materialism and Culture: Selected Essays*. London: Verso.

Woodhead, M. (1988). "When psychology informs public policy." *American Psychologist* 43(6): 443-454.

Wright, S. (1920). "The relative importance of heredity and environment in determining the piebald pattern of guinea pigs." *Proceedings of the National Academy of Science* 6: 320-332.

Index

Table Intro.1 Eight gaps in analyzing and interpreting heritability

Puzzles (Items A, C)	Gaps (Item B)	
Puzzle #3—Transfer from agricultural to human studies—motivates the examination of all the gaps.	1. *Terminology*	Key terms have multiple meanings that are distinct.
	2. *Patterns versus factors*	Statistical patterns (e.g., variation for a trait partitioned into components) are distinct from measurable underlying factors.
Puzzle #1—IQ paradox—gets partly resolved by attention to gaps 2 and 5	3. *Translation to hypotheses*	Translation from statistical analyses to hypotheses about measurable factors is difficult.
	4. *Predictions*	Predictions based on extrapolations from existing patterns of variation may not match observed outcomes.
	5. *Unreliable estimates*	Partitioning of variation, especially in human studies, does not reliably estimate the intended quantities.
Puzzle #2—Underlying heterogeneity—feeds into gap 6

More puzzles follow, especially from gaps 6 & 8 | 6. *Translation to hypotheses (in light of heterogeneity)* | Translation from statistical analyses to hypotheses about the measurable factors is even more difficult in light of the possible heterogeneity of underlying genetic or environmental factors. |
| | 7. *Causal influences* | Many steps lie between the analysis of observed traits and interventions based on well-founded claims about the causal influence of genetic and environmental factors. |
| | 8. *Within to between groups* | Explanation of variation within groups does not translate to explanation of differences among groups. |

(See Introduction and Part I for elaboration)

Appropriate Responses (Item B)	Implications (and Background) (Items D-K & book as a whole)
Meanings need to be kept distinct, for which terminological changes can help.	
Needs to be highlighted and kept open.	Blurring this distinction leads to interpretations of components of variation that are unjustified -> Say *no* to nature-nurture (NNN). Critiques of heritability studies often abet this blurring (Item H).
The steps and conditions to bridge (or circumvent) this gap warrant attention.	Heritability studies are unreliable basis for molecular or sociological investigations of traits (NNN). Restrict attention to relatives (-> Item K).
Compensate for the discrepancies (especially if any actions depend on the predictions).	Causal claims circumscribed as rerun predictability (Item F), which make sense in relation to the actions based on the basis of what is known (Item G), a theme informing Item I.
Collect data sets needed to remedy the gap. Acknowledge when this is difficult.	Further unjustified interpretations of components of variation follow (NNN). (Demonstrated by examining alternatives to three standard assumptions, Item E, aided by gene-free quantitative genetics, Item D.)
The steps and conditions to bridge this gap—or to circumvent it—warrant attention.	Even less reliable basis for molecular or sociological investigations of traits (NNN). Pay attention to possible heterogeneity of underlying factors in fields other than heritability studies, e.g., epidemiology and social sciences (Items I & J).
Recognize that estimates of heritability and other fractions of the variation are of even more limited utility than gaps 3–6 indicate.	
Recognize that this gap is firm and its implications are deep.	

www.ingramcontent.com/pod-product-compliance
Lightning Source LLC
Chambersburg PA
CBHW081523220326

41598CB00036B/6313